Biostatistics and Computer-based Analysis
of Health Data using SAS

**Biostatistics and Health Science Set**

coordinated by
Mounir Mesbah

# Biostatistics and Computer-based Analysis of Health Data using SAS

Christophe Lalanne
Mounir Mesbah

First published 2017 in Great Britain and the United States by ISTE Press Ltd and Elsevier Ltd

ISTE Press Ltd
27-37 St George's Road
London SW19 4EU
UK

www.iste.co.uk

Elsevier Ltd
The Boulevard, Langford Lane
Kidlington, Oxford, OX5 1GB
UK

www.elsevier.com

**Notices**

Knowledge and best practice in this field are constantly changing. As new research and experience broaden our understanding, changes in research methods, professional practices, or medical treatment may become necessary.

Practitioners and researchers must always rely on their own experience and knowledge in evaluating and using any information, methods, compounds, or experiments described herein. In using such information or methods they should be mindful of their own safety and the safety of others, including parties for whom they have a professional responsibility.

To the fullest extent of the law, neither the Publisher nor the authors, contributors, or editors, assume any liability for any injury and/or damage to persons or property as a matter of products liability, negligence or otherwise, or from any use or operation of any methods, products, instructions, or ideas contained in the material herein.

For information on all our publications visit our website at http://store.elsevier.com/

British Library Cataloguing-in-Publication Data
A CIP record for this book is available from the British Library
Library of Congress Cataloging in Publication Data
A catalog record for this book is available from the Library of Congress
ISBN 978-1-78548-111-6

Printed and bound in the UK and US

# Contents

# Introduction

A large number of the actions performed by means of statistical software amount to manipulating, or even to literally transforming digital data representing statistical data. It is therefore paramount to understand how statistical data are represented and how they can be used by software such as SAS. After the importing, recoding and eventual transformation of these data, the description of the variables of interest and the summary of their distribution in numerical and graphical form constitute a fundamental preparatory stage to any statistical modeling, hence the importance of these early stages in the progress of a project for statistical analysis. Secondly, it is essential to fully control the commands that enable the calculation of the main measures of association in medical research and to know how to implement the conventional explanatory and predictive models: analysis of variance, linear and logistic regression, and the Cox model. Unlike common practice with the R language, up to a few exceptions, making use of the SAS commands available on installation of the software (basic commands, procedures) will be preferred over the usage of specialized libraries of commands.

This book assumes that the reader is already familiar with basic statistical concepts, in particular the calculation of central tendency and dispersion indicators for a continuous variable, contingency tables, analysis of variance and conventional regression models. The objective here is to apply this knowledge to data sets described in numerous other works, even if the interpretation of the results remains minimal, in order to quickly familiarize oneself with the use of SAS with real data. Emphasis is particularly given to the management and the manipulation of structured data since it can be noted that this constitutes 60 to 80 % of the work of the statistician. There are many books on SAS, covering both the technical and statistical points of view. Some of these books are rather general in nature [RIN 14, DEC 11], others, on the contrary, are much more specialized and address similar subjects. The purpose of this book is to allow readers to quickly familiarize themselves with SAS such that they can conduct their own analyses and continue their learning in an autonomous way in the field of medical statistics.

This book constitutes a sequel to *Biostatistics and Computer-Based Analysis of Health Data Using R*, which was published by the same authors in the same collection [LAL 16]. Every topic that relates to data organization and data exploratory analysis, in particular graphic methods, are discussed therein. In this book, the same data sets are being used to facilitate the transfer of learning of the knowledge acquired in R.

In Chapter 1, the base commands for data management with SAS will be introduced. This primarily concerns the creation and the manipulation of quantitative and qualitative variables (recoding of individual values, counting of missing observations), importing databases stored in the form of text files, as well as elementary arithmetic operations (minimum, maximum, arithmetic mean, difference, frequency, etc.). We will also examine how to store preprocessed databases in text or in SAS formats.

The objective of Chapter 2 is to understand how the data are represented in SAS and how to work with them. Commands useful for describing a data table composed of quantitative or qualitative variables are also presented therein. The descriptive approach is strictly univariate, which constitutes the prerequisite for any statistical approach. Base graphic commands (histograms, density curves, bar or dot plots) will be presented in addition to the usual central tendency (mean, median) and dispersion (variance, quartiles) numerical descriptive summaries. Pointwise and interval estimation using arithmetic means and empirical proportions will also be addressed. The objective is to become familiar with the use of simple SAS commands operating on a variable, optionally specifying certain options for the calculation, alongside the selection of statistical units among all of the available observations.

Chapter 3 is dedicated to the comparison of two samples for quantitative or qualitative measurements. The following hypothesis tests are addressed: the Student's t-test for independent or paired samples, the non-parametric Wilcoxon test, the $\chi^2$ test and the Fisher's exact test, as well as the NcNemar test based on the main measures of association for two variables (average difference, odds ratio and relative risk). From this chapter onwards, there will be less emphasis on the univariate description of each variable, but it is advisable to always carry out the stages of data description discussed in this chapter. The objective is to control the main statistical tests in the case where the relationship between a quantitative variable and a qualitative variable, or for two qualitative variables, is the main interest. This chapter also presents analysis of variance (ANOVA) where we explain the variability observed at the level of a numerical response variable by taking a group or classification factor into account, and the estimation with confidence intervals of average differences. Emphasis will be placed on the construction of an ANOVA table summarizing the various sources of variability and on the graphic methods that can be used to summarize the distribution of individual or aggregated data. The linear tendency test will also be studied when the classification factor can be considered as naturally ordered. The objective is to understand how to construct an explanatory model in the case where there is one or even two explanatory factors, and how to digitally and graphically present the results of such a model through the use of SAS.

Chapter 4 focuses on the analysis of the linear relation between two continuous quantitative variables. In the linear correlation approach, which assumes a symmetrical relation between the two variables, the main emphasis will be on quantifying the force and the direction of the association in a parametric (Pearson's correlation) or in a non-parametric manner (rank-based Spearman's correlation) and on the graphic representation of this relation. Simple linear regression will be used in the event that one of the two numeric variables assumes the function of a response variable and the other that of an explanatory variable. The useful commands for the estimation of the coefficients of the regression line, the construction of the ANOVA table associated with the regression and the computation of fitted values will be presented. The objective of this chapter remains identical to that of Chapter 3, namely to present the SAS commands necessary for the construction of a simple statistical model between two variables, in an explanatory or predictive perspective.

In Chapter 5, the main measures of association found in epidemiological studies will be discussed: odds ratio, relative risk, prevalence, etc. SAS commands enabling estimation (pointwise and by interval) and the associated hypothesis tests will be illustrated with cohort or case-control studies data. The implementation of a simple logistic regression model makes it possible to complete the range of statistical methods, allowing the observed variability to be explained at the level of binary response variables. The objective is to understand the SAS commands to be used in the case in which the variables are binary, either to summarize a contingency table in the form of association indicators or to model the relationship between a binary response (ill/healthy) and a qualitative explanatory variable based on the so-called grouped data. Two good bibliographical references are [BLI 52, BRE 80].

Chapter 6 constitutes an introduction to the analysis of censored data, the main tests associated with the construction of a survival curve (log-rank or Wilcoxon tests) and finally the Cox regression model. The specificity of the censored data requires particular care in the coding of data in SAS, and the objective is to present the SAS commands essential to the correct representation of survival data in digital form, to their numerical (survival median) and graphical (Kaplan-Meier curve) summary, and the implementation of common tests.

At the end of each chapter, a few applications are proposed and examples of commands that will help in answering the questions being raised are available for most of the questions. It is sometimes possible to obtain identical results with other approaches or by employing other commands. Outputs from SAS are not reproduced but readers are encouraged to try the proposed SAS instructions themselves and to try alternative or complementary instructions. It will be assumed that the data files being used are available in the working directory. All of the data files and the SAS commands used in this book can be downloaded from the companion website (https//github.com/biostatsante).

Due to formatting reasons, some of the SAS outputs have been truncated or reformatted. As a result, this could present differences when the reader attempts to reproduce the commands mentioned in this book.

The procedures and SAS commands utilized in the various chapters are listed at the end of each chapter. We present these procedures only in their simplest forms. The objective is to make every reader able to begin utilizing SAS. It is not possible to show all the possibilities of usage of these procedures here. The reader will discover them through practice.

# Language Elements

This chapter will mainly focus on the way in which data are represented in SAS and on their manipulation, by operating on subsets of variables or by only selecting certain observations.

## 1.1. Introduction to the SAS language

### 1.1.1. *What's SAS?*

SAS is a software program which was developed in the early 1970s at the *North Carolina State University at Chapel Hill*, USA. Undoubtedly today, it is by far the most commonly utilized statistical software.

It stems from an acronym for "**S**tatistical **A**nalysis **S**ystem" and is pronounced *SASSE*, rather than *ES A ES*.

SAS comprises different modules:

– **SAS/Base:** data management and basic procedures;

– **SAS/STAT:** statistical analysis;

– **SAS/GRAPH:** graphics;

– **SAS/OR:** operational research;

– **SAS/ETS:** econometrics and time series;

– **SAS/IML:** interactive matrix language;

– **SAS/QC:** quality control.

Only the first three modules will be covered in this book.

### 1.1.2. *Structure and basic elements of the SAS language*

SAS programs must be written in an EDITOR window designed for this purpose. The execution of a SAS program is carried out by selecting it and clicking on the little man in the taskbar. It is possible to suspend or stop a program. Results are displayed in a RESULTS window. Notes about this execution are displayed in the LOG window. It is strongly recommended to always read them before checking the results.

Most SAS programs are mainly constituted of:

1) DATA steps, to read the data, transform them, prepare them for the next stage (in the following stages) where they are analyzed by SAS procedures;

2) PROC steps, to analyze the data with statistical procedures from the SAS Base module (descriptive, etc.), or more in-depth, from the SAS STAT module.

A SAS program consists of a set of statements (instructions) each ending with a semicolon (;). A new line (line feed) is not enough to complete the statement.

Within a statement, the slash (/) symbol usually separates standard usage from more options being added.

Each SAS program ends with: **run ;**.

The program is executed only after reading a **run ;**.

SAS essentially works with "rectangular" files, commonly called tables, which will contain the data. The table names cannot begin with a number. The tables contain observations, which are the table rows, and variables, the table columns.

The variable _n_ contains the observation number in the SAS table.

Missing data are coded by a period (.). It is possible to encode them in some other manner.

The name of the SAS variables can comprise up to 32 characters, letters, numbers and _. They cannot begin with a number. Avoid beginning with a _ (underscore symbol) because some SAS system variables often start this way. There is no difference between upper case and lower case. White spaces have no effect, except inside parentheses or between quotes.

Comments written between slash star (/*) and star slash */ or between a star (*) and a semicolon (;) are not compiled during execution. They appear in green in the editor screen.

Basic operators can be used in some instructions, essentially in data steps, but not exclusively.

1) comparison:

- = equal,

- ^= not equal,

- < or lt: less than,

- > or gt: greater than,

- < = or le: less or equal,

- >= or ge: greater or equal;

2) logic: and, or, if, then, else, do, else, end.

## 1.2. Creating and managing SAS tables

### 1.2.1. *Creating a SAS table*

This little SAS program can be written to create a SAS table containing two variables of the alphabetical type (character string), name and gender, and three numeric variables, age, weight and size.

```
DATA tablename;
INPUT name $ gender $ age weight size;
CARDS;
Marie M 23 47 1.50
Mohamed F 33 80 1.73
Mamadou M 50 75 1.66
Francis M 58 70 1.75
; RUN;
```

The reading format is free. Upon execution of this program, the log window will display:

```
NOTE: The table WORK.TABLENAME has 4 observations and
  5 variables.
NOTE: The DATA statement used (Total process time):
real time 0.06 seconds
cpu time 0.00 seconds
```

In the SAS explorer, inside the "Work" directory, the file "tablename" will be created which can be opened by double clicking on it.

| | name | gender | age | weight | size |
|---|---|---|---|---|---|
| 1 | Marie | M | 23 | 47 | 1.5 |
| 2 | Mohamed | F | 33 | 80 | 1.73 |
| 3 | Mamadou | M | 50 | 75 | 1.66 |
| 4 | Francis | M | 58 | 70 | 1.75 |

**Figure 1.1.** *Preview of the file "tablename"*

If these data were in a text file, named "file.txt", located in the directory "c:\folder\subfolder\", there would be no need to retype them in the Editor window. The program is thus simplified:

```
DATA tablename;
INFILE 'c:\folder\subfolder\file.txt';
INPUT name $ gender $ age weight size;
RUN;
```

The result will be identical to the previous one.

### 1.2.2. *Creating a SAS table from another one*

These three programs will give the same result: they allow the creation of a new SAS table based on the previous one and keeping only the variables age and gender.

```
DATA tablename2 (keep = age gender);
SET tablename;
RUN;

DATA tablename2;
SET tablename (keep = age gender);
RUN;

DATA tablename2;
SET tablename;
keep age gender;
RUN;
```

In the program that follows, the newly created table does not contain the variable "age",

```
DATA tablename2;
SET tablename1;
DROP age;
RUN;
```

### 1.2.3. *Creating permanent SAS tables*

```
LIBNAME tot 'c:\folder\subfolder\';
DATA tot.table1;
X=25;
Y=X*2;
RUN;
```

The following notes will appear in the log window:

```
59 LIBNAME tot 'c:\folder\subfolder\';
NOTE: Libref TOT was successfully assigned as follows:
Engine: V9
Physical name: c:\folder\subfolder\
60 DATA tot.table1;
61 X=25;
62 Y=X*2;
63 RUN;
```

```
NOTE: The table TOT.TABLE1 has 1 observations and 2 variables.
NOTE: The DATA step has used (Total process time):
```

You will not see the table in the "Work" directory of the SAS Explorer. In the SAS environment, a directory called "tot", which will contain the table "table1", will appear inside the Explorer window.

In addition, and above all, table "table1" will be created permanently in your Windows environment (or other, when utilizing another system), inside the directory specified by the path indicated, under the name of: table1.sas7bcat. It is created on a permanent basis, unlike previous ones, which will disappear when leaving the SAS environment. It is possible to export it.

### 1.2.4. *Importing tables of any format*

The PROC IMPORT procedure and its interactive interface. Click on the File menu at the top left, then on the sub-menu import data and follow the guidelines to import data in various formats existing in your Windows environment.

The following file formats can be imported:

Text, Excel, Access, csv, dbf (DBASE), jmp (JMP), sav (SPSS), dta (Stata), db (paradox) and wk (lotus).

In addition, it is possible to have a copy of the PROC IMPORT program, which enables this import to take place.

The following example shows how to import the file "birthwt.txt" [HOS 89], located in the "subfolder" sub-directory of the "folder" directory of our Windows environment. This is a text file consisting of 10 columns, corresponding to 10 variables, and of 189 rows, corresponding to 188 individuals and a first row containing the names of the variables.

```
PROC IMPORT OUT= WORK.BIRTHWT
DATAFILE="C:\folder\subfolder\birthwt.txt"
DBMS=TAB REPLACE;
GETNAMESYES;
DATAROW=2;
RUN;
```

The preview of the first observations is presented in Figure 1.2 below.

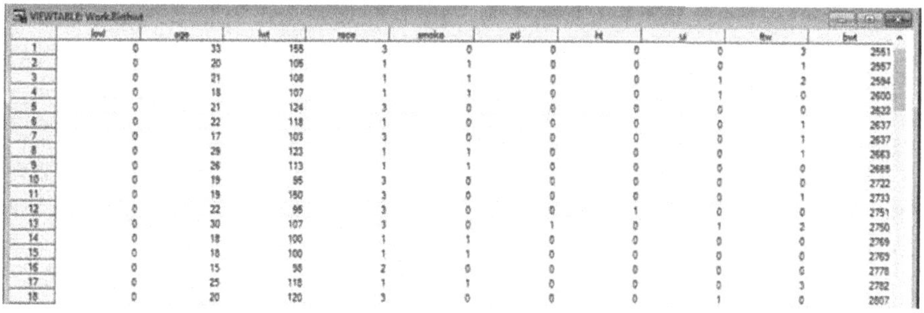

| | low | age | lwt | race | smoke | ptl | ht | ui | ftv | bwt |
|---|---|---|---|---|---|---|---|---|---|---|
| 1 | 0 | 33 | 155 | 3 | 0 | 0 | 0 | 0 | 3 | 2551 |
| 2 | 0 | 20 | 105 | 1 | 1 | 0 | 0 | 0 | 1 | 2557 |
| 3 | 0 | 21 | 108 | 1 | 1 | 0 | 0 | 1 | 2 | 2594 |
| 4 | 0 | 18 | 107 | 1 | 1 | 0 | 0 | 1 | 0 | 2600 |
| 5 | 0 | 21 | 124 | 3 | 0 | 0 | 0 | 0 | 0 | 2622 |
| 6 | 0 | 22 | 118 | 1 | 0 | 0 | 0 | 0 | 1 | 2637 |
| 7 | 0 | 17 | 103 | 3 | 0 | 0 | 0 | 0 | 1 | 2637 |
| 8 | 0 | 29 | 123 | 1 | 1 | 0 | 0 | 0 | 1 | 2663 |
| 9 | 0 | 26 | 113 | 1 | 1 | 0 | 0 | 0 | 0 | 2665 |
| 10 | 0 | 19 | 95 | 3 | 0 | 0 | 0 | 0 | 0 | 2722 |
| 11 | 0 | 19 | 150 | 3 | 0 | 0 | 0 | 0 | 1 | 2733 |
| 12 | 0 | 22 | 95 | 3 | 0 | 0 | 1 | 0 | 0 | 2751 |
| 13 | 0 | 30 | 107 | 3 | 0 | 1 | 0 | 1 | 2 | 2750 |
| 14 | 0 | 18 | 100 | 1 | 1 | 0 | 0 | 0 | 0 | 2769 |
| 15 | 0 | 18 | 100 | 1 | 1 | 0 | 0 | 0 | 0 | 2769 |
| 16 | 0 | 15 | 98 | 2 | 0 | 0 | 0 | 0 | 0 | 2778 |
| 17 | 0 | 25 | 118 | 1 | 1 | 0 | 0 | 0 | 3 | 2782 |
| 18 | 0 | 20 | 120 | 3 | 0 | 0 | 0 | 1 | 0 | 2807 |

**Figure 1.2.** *Preview of the imported SAS file "Birthwt"*

With the following SAS Proc FORMAT procedure, we can clarify some of the codings in the following text. The modalities of some of the categorical variables can be defined by making use, in certain procedures, of the formats defined by PROC FORMAT.

A large number of SAS procedures recognize the labels defined by PROC FORMAT.

Starting from the following section, we will see simple examples where these labels are employed.

```
PROC FORMAT;
value low 1="Weight less than 2.5 Kg"
         0="Weight greater than 2.5 Kg";
value ethnicity 1="White"
2="Black"
3="Other";
value tabac 1="Tobacco consumption during pregnancy"
0="No tobacco consumption during pregnancy";
value Hypert 1="History of hypertension"
0="No hypertension history";
value uterine 1="Manifestation of uterine irritability"
0="No manifestation of uterine irritability";
run;
```

### 1.2.5. *Accessing the contents of a SAS table*

```
PROC CONTENTS data= tablename;
RUN;
```

The following information is generated in the table: number of observations, number of variables, types and variable labels (when these labels exist), etc.

For the previous example, we will see in the results window:

The CONTENTS Procedure

| Data Set Name | WORK.TABLENAME | Observations | 4 |
|---|---|---|---|
| Member Type | DATA | Variables | 5 |
| Engine | V9 | Indexes | 0 |
| Created | 23/01/2017 16:11:54 | Observation Length | 40 |
| Last Modified | 23/01/2017 16:11:54 | Deleted Observations | 0 |
| Protection | | Compressed | NO |
| Data Set Type | | Sorted | NO |
| Label | | | |
| Data Representation | WINDOWS_64 | | |
| Encoding | wlatin1 Western (Windows) | | |

**Figure 1.3.** *Results from the execution of the CONTENTS procedure*

## 1.2.6. *Visualizing the contents of SAS tables*

```
PROC PRINT data= tablename;
run;
```

| Engine/Host Dependent Information | |
|---|---|
| Data Set Page Size | 65536 |
| Number of Data Set Pages | 1 |
| First Data Page | 1 |
| Max Obs per Page | 1632 |
| Obs in First Data Page | 4 |
| Number of Data Set Repairs | 0 |
| ExtendObsCounter | YES |
| Filename | C:\Users\MOUNIR~1\AppData\Local\Temp\SAS Temporary Files\_TD1852_FUJLIFEBOOK_\tablen: |
| Release Created | 9.0401M2 |
| Host Created | X64_8PRO |

| Alphabetic List of Variables and Attributes | | | |
|---|---|---|---|
| # | Variable | Type | Len |
| 3 | age | Num | 8 |
| 2 | gender | Char | 8 |
| 1 | name | Char | 8 |
| 5 | size | Num | 8 |
| 4 | weight | Num | 8 |

**Figure 1.4.** *Results from the execution of the CONTENTS procedure, continuation*

| Obs | name | gender | age | weight | size |
|---|---|---|---|---|---|
| 1 | Marie | M | 23 | 47 | 1.50 |
| 2 | Mohamed | F | 33 | 80 | 1.73 |
| 3 | Mamadou | M | 50 | 75 | 1.66 |
| 4 | Francis | M | 58 | 70 | 1.75 |

When considering only certain variables of the table:

```
PROC PRINT data= tablename;
VAR name age;
RUN;
```

| Obs | name | age |
|---|---|---|
| 1 | Marie | 23 |
| 2 | Mohamed | 33 |
| 3 | Mamadou | 50 |
| 4 | Francis | 58 |

When considering only certain observations of the table:

```
Data;
set tablename;
if _n_=2;
PROC PRINT ;
run;
```

Obs name gender age weight size

1 Mohamed F 33 80 1.73

We can also use a data step to modify or correct observation values:

```
Data tablename; set tablename;
If name="Marie" then gender="F";
If name="Mohamed" then gender="M";
Run;
```

### 1.2.7. *Creation and transformation of variables in a SAS table*

In a data stage, a large number of basic mathematical functions are available, the most common being:

Napierian logarithm: LOG(<number>)

Exponential: EXP(<number>)

Power: **(<number>)

Square root: SQRT(<number>)

Absolute value: ABS(<number>)

Integer part: INT(<number>)

There are also more complex functions available: random number generation (with specific distributions), statistical functions, etc.

In addition, to create new variables, from those existing in the old file, we can use the logical operators and, or as well as

IF <condition> THEN <instruction A>; ELSE <instruction B>;

In the example that follows,

```
DATA tablename4;
SET tablename1;
BMI = WEIGHT/(SIZE*SIZE);
RUN;
```

The file "tablename4" contains a new variable "BMI", created from the variables "weight" and "size".

In the following example:

```
DATA tablename5;
SET tablename1;
if age le 45 then age_c=1;
else age_c=2;
RUN;
```

A variable "age_c" is created, it is coded as 1 or 2 depending on whether the individual's age is less than or equal to 45 or not.

### 1.2.8. *Creation of labels with PROC FORMAT and their usage*

We can create labels with the procedure Proc Format

```
PROC FORMAT;
VALUE BG 0 = 'Bad' 1 = 'Good' -1 = 'Missing';
VALUE Genre 1 = 'Male' 2 = 'Female' 0 = 'Missing';
VALUE AGeCl LOW - 35 = 'Young' 35 -HIGH ='Old' ;
RUN;
```

Observations: once these labels created by PROC FORMAT (that is, PROC FORMAT is executed), they can be assigned to a variable within a data stage:

```
DATA new;
SET tablename;
format age ageCL.;
run;
```

Individuals under the age of 35 will have the label "young", others the label "old". In the file "new", the original ages are permanently deleted and replaced with the labels. Therefore, the execution of the following program

```
PROC PRINT data= new;
run;
```

will give the listing:

| Obs | name | gender | age | weight | size |
|-----|------|--------|-----|--------|------|
| 1 | Marie | F | Young | 47 | 1.50 |
| 2 | Mohamed | M | Young | 80 | 1.73 |
| 3 | Mamadou | M | Old | 75 | 1.66 |
| 4 | Francis | M | Old | 70 | 1.75 |

Another way of proceeding, after running PROC FORMAT, is to use the instruction "format" in PROC PRINT, applied to the file "tablename":

```
PROC PRINT data= tablename; format age agecl.; run;
```

Therefore, without modifying the data file, we can use PROC FORMAT to recode variables.

### 1.2.9. *Sorting SAS tables*

```
PROC SORT data = tot.table1;
BY Age;
Run;
```

The table "tot.table1" is sorted by increasing age. The new sorted table overrides the former.

```
PROC SORT data=tot.table1
OUT=tot.table3;
BY gender;
RUN;
```

The table "tot.table1" is sorted by increasing gender (alphabetical, if "gender" is a variable of the alphanumeric type). The new sorted table is stored under the name "tot.table3". The former table "table1" is preserved.

```
PROC SORT data=tot.table1 NODUPKEY;
BY number;
RUN;
```

The table "tot.table1" is sorted by increasing number. Duplicates are eliminated. To be used with care.

**BY statement:**

BY variablename;

This is a statement that can be employed with any procedure. It makes it possible to execute the procedure sequentially for all the observed modalities/values of the variable in the table. The table is divided into several tables corresponding to each observed modality/value, and then the procedure is separately run in each of these sub-tables. It is nonetheless essential that the table is previously sorted based on this variable (using PROC SORT ; BY variablename ;).

### 1.2.10. *Concatenating SAS tables*

The following example shows how to concatenate tables "tot.table1" and "tot.table2" into a new table, "table3". Note the essential use of the instruction **BY**.

```
PROC SORT data = tot.table1;BY number;
RUN;
```

The table "tot.table1" is sorted by increasing number.

```
PROC SORT data = tot.table2;BY number;
RUN;
```

The table "tot.table2" is sorted by increasing number.

```
DATA data =table3;MERGE tot.table1 tot.table2};
BY number;
RUN;
```

The new table, table3 is created. Table3 contains for each value of the variable "number", the variables of "tot.table1" and of "tot.table2". It is advisable to always verify that the table obtained is the desired one.

## 1.3. Key points to remember

– SAS represents data as a list of variables, similar to a data table in which the variables are arranged in columns, all having the same number of observations, and the values of the variables are usually numbers that can be associated with labels when they refer to the modalities of a qualitative variable.

– The creation, manipulation and management of these SAS tables are usually done in DATA steps.

– The analysis of these data is done in PROC steps (for procedure).

– In this chapter, we have introduced the basic elements of the SAS language. Examples of simple programs are presented including DATA steps and various basic procedures: PROC IMPORT (to import tables created in environments other than SAS), PROC CONTENTS (to display a list of variables in a table), PROC PRINT (to print the contents of a table), PROC FORMAT (to create labels), PROC SORT (to sort tables), PROC FREQ (presented in applications, allows simple or cross tabulations of qualitative variables to be displayed) and PROC SUMMARY (presented in applications, making it possible to display simple or cross tabulations of quantitative variables).

## 1.4. Further information

SAS is a renowned piece of software for facilitating database management. For more details, S. Ringuede's manual [RIN 14] or the SAS Base reference manual [SAS 15], available as a PDF, provides a detailed description of the set of controls related to the management and manipulation of variables.

PROC SQL is an advanced SAS procedure, dedicated to database management. The latest edition of S. Ringuede's manual [RIN 14] addresses a chapter thereto.

Finally, for all matters regarding aspects related to automation or programming with SAS, the reader is invited to consult the appendices: SAS IML and SAS MACRO or the second edition of S. Tuffery's book [TUF 11].

## 1.5. Applications

1) A researcher has collected the following biological measurements (arbitrary units):

3.68  2.21  2.45  8.64  4.32  3.43  5.11  3.87

a) Store the sequence of measurements in a variable called x.

b) Indicate the number of observations, the minimal and maximal values, as well as the range.

c) In fact, the researcher realizes that the value 8.64 corresponds to an input error and should be changed to 3.64. Similarly, he has doubts about the 7th measurement and decides that it should be considered as a missing value: perform the corresponding transformations.

To capture data in SAS, a DATA step is initially carried out in which the individual data are manually input using the command "input" to indicate the name of the variable, followed by cards that put the data directly inside the SAS program.

```
DATA Exercise1_1;
INPUT x;
CARDS;
3.68
2.21
2.45
8.64
4.32
3.43
5.11
3.87
;
RUN;
```

The number of observations as well as the values of the extreme observations of the empirical distribution are obtained from the command

PROC SUMMARY, by specifying the corresponding options (n, min and max, and range).

```
PROC SUMMARY  DATA=Exercise1_1  PRINT  n   min   max   range;
VAR x;
RUN;
```

Given that the data set is limited in terms of values, the easiest way consists of modifying the previous DATA step by performing the requested modifications: (i) replacing the value 8.64 by 3.64 and (ii) recoding the value 5.11 as a point (" . ") which is the default symbol used by SAS to represent missing data.

```
DATA Exercise1_1; INPUT x;
CARDS;
3.68
2.21
2.45
3.64           /*  (a)  */
4.32
3.43
.        /*  (b)  */
3.87
; RUN;
```

2) The plasma viral load provides a means to describe the quantity of a virus (e.g. HIV) in a blood sample. This virological marker, which makes it possible to track the progress of the infection and to measure the effectiveness of treatments, is expressed as the number of copies per milliliter, and most measurement instruments have a detectability threshold of 50 copies/ml. Here follows a series of measures, $X$, expressed in logarithms (base 10) collected from 20 patients:

3.64 2.27 1.43 1.77 4.62 3.04 1.01 2.14 3.02 5.62 5.51 5.51
1.01 1.05 4.19 2.63 4.34 4.85 4.02 5.92

As a reminder, a viral load of $100,000$ copies/ml is equivalent to 5 log.

a) Indicate how many patients have a viral load considered as non-detectable.

b) What is the median viral load level, in copies/ml, for the data to be considered as valid?

Data input is achieved in the same way as in the previous exercise, that is, based on a DATA step.

```
DATA Exercise1_2;
INPUT  X;
detect=1;
IF  X  <=  log10(50)  THEN detect=0; CARDS;
3.64
2.27
1.43
1.77
4.62
3.04
1.01
2.14
3.02
5.62
5.51
5.51
1.01
1.05
4.19
2.63
4.34
4.85
4.02
5.92
; RUN;
```

A small difference should be noted compared to example 1, namely that a "detect" variable is also defined which takes the value 0 if X = log10(50), and 1 otherwise. This is performed before the instruction cards which designate the start of storing the numerical data into memory. The number of patients with an undetectable viral load is obtained from a simple frequency tabulation of the variable "detect" by means of the command PROC FREQ.

```
PROC FREQ  DATA=exercise1 2;  TABLES  detect;  RUN;
```

To calculate the median viral load for the valid observations only (that is, having a value above the detection threshold limit), it is necessary to select the statistical units satisfying the validity conditions and to use the command PROC SUMMARY.

```
DATA detect; SET   exercise1_2;
Y=exp(X*log(10));
IF  detect=1;
RUN;
PROC SUMMARY DATA=detect   PRINT   median;
VAR Y;
RUN;
```

3) The file anorexia.dat contains data from a clinical study among anorexic patients who received one of the three following therapies: behavioral therapy, family therapy and control therapy [HAN 93].

a) How many patients are there in total? How many patients are there per treatment group?

b) Weight measurements are in pounds. Convert them into kilograms.

c) Create a new variable containing the difference scores (After - Before).

To read the data contained in the file anorexia.data, of which a preview is provided below:

```
Group Before After
g1 80.5   82.2
g1 84.9   85.6
g1 81.5   81.4
g1 82.6   81.9
g1 79.9   76.4
```

we are going to employ the command "infile" in the DATA step, specifying that reading the data should start at the 2nd row (firstobs=2) and that the data are separated by spaces (in varying numbers).

```
DATA anorexia;
INFILE   "C:\data\anorexia.dat"   firstobs=2   dlm="09"X;
INPUT      group $  1-2      before  3-7 after  8-13;
RUN;
```

We can very well associate more informative labels with the procedures taken by the qualitative variable "group" by means of a PROC FORMAT:

```
PROC FORMAT; VALUE Therapy
1='Behavioral therapy'
2='Family therapy'
3='Control therapy'
;
run;
```

In order to obtain the values per therapy type, we use a simple frequency tabulation by means of the command PROC FREQ. The displaying of the results can be customized by utilizing the renaming of the groups obtained in the previous step.

```
PROC FREQ  DATA=anorexia;  TABLES  group;
FORMAT group therapy.;
RUN;
```

The transformation of the weight units does not cause any specific problem, but we should decide whether new variables have to be created or existing values must be replaced. Here, we will create two new variables, "before_kg" and "after_kg":

```
DATA anorexia;  SET  anorexia;
before_kg=before/2.2; after_kg=after/2.2;
RUN;
```

For the difference values, a new variable is also created by means of a DATA step:

```
DATA anorexia;  SET  anorexia;
diff=after_kg-before_kg;
RUN;
```

# 2

# Simple Descriptive Statistics

In this chapter, we will look at the main commands capable of summarizing, in a quantitative or graphical manner, the distribution of a numeric or categorical variable.

Generally, including the ODS GRAPHICS ON statement before each procedure, by default, results in some graphic outputs illustrating the bivariate or univariate distributions described by these procedures. In some SAS environments, it is active (ON) by default. The ODS GRAPHICS OFF statement makes it possible to disable it.

## 2.1. Univariate descriptive statistics: Estimation

### 2.1.1. *Proportions, simple tabulation*

The most commonly used procedure is the PROC FREQ procedure. Other procedures can yield similar results, at least from a descriptive perspective (Proc tabulate allows reporting).

```
PROC FREQ data=birthwt;
TABLES race smoke;
format race ethnicity. smoke tobacco.;
RUN;
```

The results obtained are:

The example below illustrates the use of the BY statement.

```
PROC SORT data= birthwt;
BY race;
RUN;
PROC FREQ data=birthwt;
TABLES smoke ;
format race ethnicity. smoke tobacco.;
BY race;
RUN;
```

The FREQ Procedure

| race | Frequency | Percent | Cumulative Frequency | Cumulative Percent |
|------|-----------|---------|----------------------|--------------------|
| White | 96 | 51.06 | 96 | 51.06 |
| Black | 25 | 13.30 | 121 | 64.36 |
| Other | 67 | 35.64 | 188 | 100.00 |

| smoke | Frequency | Percent | Cumulative Frequency | Cumulative Percent |
|-------|-----------|---------|----------------------|--------------------|
| No tobacco consumption during pregnancy | 114 | 60.64 | 114 | 60.64 |
| Tobacco consumption during pregnancy | 74 | 39.36 | 188 | 100.00 |

**Figure 2.1.** *Results from the execution of the FREQ procedure*

The FREQ Procedure

race=White

| smoke | Frequency | Percent | Cumulative Frequency | Cumulative Percent |
|-------|-----------|---------|----------------------|--------------------|
| No tobacco consumption during pregnancy | 44 | 45.83 | 44 | 45.83 |
| Tobacco consumption during pregnancy | 52 | 54.17 | 96 | 100.00 |

**Figure 2.2.** *Results from the execution of the FREQ procedure using the BY statement*

## 2.1.2. *Quantitative variables*

```
PROC SUMMARY data= birthwt PRINT mean stddev min max median;
VAR age;
RUN;
```

The FREQ Procedure

race=Black

| smoke | Frequency | Percent | Cumulative Frequency | Cumulative Percent |
|---|---|---|---|---|
| No tobacco consumption during pregnancy | 15 | 60.00 | 15 | 60.00 |
| Tobacco consumption during pregnancy | 10 | 40.00 | 25 | 100.00 |

race=Other

| smoke | Frequency | Percent | Cumulative Frequency | Cumulative Percent |
|---|---|---|---|---|
| No tobacco consumption during pregnancy | 55 | 82.09 | 55 | 82.09 |
| Tobacco consumption during pregnancy | 12 | 17.91 | 67 | 100.00 |

**Figure 2.3.** *Continuation of the results from the execution of the FREQ procedure using the BY statement*

The SUMMARY Procedure

| Analysis Variable : age | | | | |
|---|---|---|---|---|
| Mean | Std Dev | Minimum | Maximum | Median |
| 23.2606383 | 5.3037313 | 14.0000000 | 45.0000000 | 23.0000000 |

**Figure 2.4.** *Results from the execution of the SUMMARY procedure*

The same result would have been obtained with PROC MEANS:

```
PROC MEANS data= birthwt PRINT mean stddev min max median;
VAR age;
RUN;
```

The MEANS Procedure

| Analysis Variable : age | | | | |
|---|---|---|---|---|
| Mean | Std Dev | Minimum | Maximum | Median |
| 23.2606383 | 5.3037313 | 14.0000000 | 45.0000000 | 23.0000000 |

**Figure 2.5.** *Results from the execution of the MEANS procedure*

These two very similar procedures are nonetheless different: namely PROC MEANS will output results by default, even without the PRINT option, which is not the case of PROC SUMMARY.

Therefore, executing the following program yields:

```
PROC SUMMARY data= birthwt;/* Without PRINT option */
VAR age;
RUN;
```

will not output anything in the results window and will give in the log, the error:

ERROR: neither a PRINT option nor a valid output statement has been indicated.

Whereas executing the equivalent program with PROC MEANS yields:

```
PROC MEANS data= birthwt ; /* Without PRINT option */
VAR age;
RUN;
```

will give in the results window:

The MEANS Procedure

| Analysis Variable : age | | | | |
|---|---|---|---|---|
| N | Mean | Std Dev | Minimum | Maximum |
| 188 | 23.2606383 | 5.3037313 | 14.0000000 | 45.0000000 |

Figure 2.6. *Results from the execution of the MEANS procedure without PRINT*

For PROC MEANS, the value, the mean, the standard deviation, the minimum and the maximum are therefore given by default.

The following example illustrates a SAS statement, CLASS, which avoids the use of BY. The same program works with PROC SUMMARY.

```
PROC MEANS data= birthwt PRINT mean stddev ;
CLASS smoke;
VAR age;
format smoke tobacco.;
RUN;
```

**The MEANS Procedure**

| Analysis Variable : age | | | |
|---|---|---|---|
| smoke | N Obs | Mean | Std Dev |
| No tobacco consumption during pregnancy | 114 | 23.4649123 | 5.4759003 |
| Tobacco consumption during pregnancy | 74 | 22.9459459 | 5.0474241 |

**Figure 2.7.** *Results from the execution of the MEANS procedure with CLASS*

PROC UNIVARIATE is another concurrent procedure. By default, it gives more details about the distribution of the variables and allows for very helpful graphical representations for testing the normality of the variable, for example.

### 2.1.3. *Student's t-test*

The following example will produce the results of the Student's t-test with independent series.

```
PROC TTEST data= birthwt;
CLASS smoke;
VAR age;
format smoke tobacco.;
RUN;
```

Hereafter, the results from comparing variances are given (in this case, they do not significantly differ) as well as the result of the test in both cases: whether these variances be different or not. The Satterthwaite approximation is also given, it can be used in the case where variances are significantly different.

The TTEST Procedure

Variable: age

| smoke | N | Mean | Std Dev | Std Err | Minimum | Maximum |
|---|---|---|---|---|---|---|
| No tobacco consumption during pregnancy | 114 | 23.4649 | 5.4759 | 0.5129 | 14.0000 | 45.0000 |
| Tobacco consumption during pregnancy | 74 | 22.9459 | 5.0474 | 0.5868 | 14.0000 | 35.0000 |
| Diff (1-2) | | 0.5190 | 5.3119 | 0.7930 | | |

| smoke | Method | Mean | 95% CL Mean | | Std Dev | 95% CL Std Dev | |
|---|---|---|---|---|---|---|---|
| No tobacco consumption during pregnancy | | 23.4649 | 22.4488 | 24.4810 | 5.4759 | 4.8456 | 6.2962 |
| Tobacco consumption during pregnancy | | 22.9459 | 21.7766 | 24.1153 | 5.0474 | 4.3449 | 6.0231 |
| Diff (1-2) | Pooled | 0.5190 | -1.0454 | 2.0833 | 5.3119 | 4.8225 | 5.9126 |
| Diff (1-2) | Satterthwaite | 0.5190 | -1.0197 | 2.0577 | | | |

| Method | Variances | DF | t Value | Pr > |t| |
|---|---|---|---|---|
| Pooled | Equal | 186 | 0.65 | 0.5136 |
| Satterthwaite | Unequal | 164.95 | 0.67 | 0.5064 |

| Equality of Variances | | | | |
|---|---|---|---|---|
| Method | Num DF | Den DF | F Value | Pr > F |
| Folded F | 113 | 73 | 1.18 | 0.4562 |

**Figure 2.8.** *Results from the execution of the TTEST procedure*

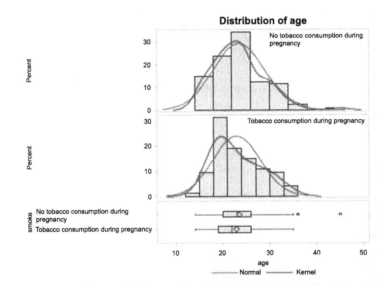

**Figure 2.9.** *Continuation of the results from the execution of the TTEST procedure. For a color version of this figure, see www.iste.co.uk/lalanne/biostatistics3.zip*

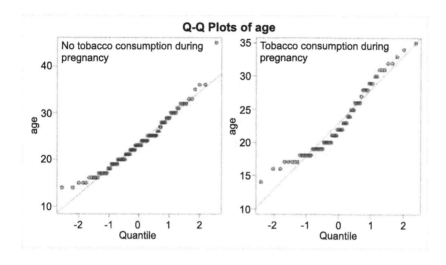

**Figure 2.10.** *Continuation of the results from the execution of the TTEST procedure*

The graphs of the histograms are also provided here by default, for the quantitative variable in each class of the group variable, as well as the densities estimated under a Gaussian (normal) or non-parametric (kernel) model.

Finally, the graphics of the QQ plots of the quantitative variable are given by class of the group variable.

The syntax of the Student's t-test with paired series is as follows. In the following example, where Lwt and Bwt are both paired series, the code will be:

```
PROC TTEST data= birthwt;
PAIRED lwt*bwt;
run;
```

### 2.1.4. *Graphics*

```
PROC Gchart data= birthwt;
HBAR age;
RUN;
```

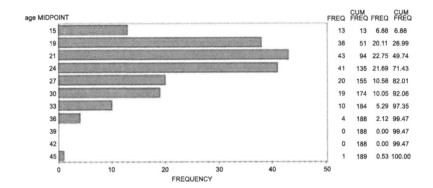

**Figure 2.11.** *Horizontal histogram of the variable age*

```
PROC Gchart data= birthwt;
VBAR age;
RUN;
```

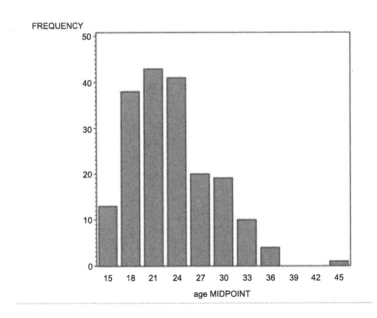

**Figure 2.12.** *Horizontal histogram of the variable age*

## 2.2. Bivariate descriptive statistics

### 2.2.1. *Contingency tables*

```
PROC FREQ data= birthwt;
TABLES smoke*race;
format race ethnicity. smoke tobacco.;
RUN;
```

This program will give the following result:

**The FREQ Procedure**

| Frequency Percent Row Pct Col Pct | Table of smoke by race | | | |
|---|---|---|---|---|
| | | race | | |
| smoke | White | Black | Other | Total |
| No tobacco consumption during pregnancy | 44 23.40 38.60 45.83 | 15 7.98 13.16 60.00 | 55 29.26 48.25 82.09 | 114 60.64 |
| Tobacco consumption during pregnancy | 52 27.66 70.27 54.17 | 10 5.32 13.51 40.00 | 12 6.38 16.22 17.91 | 74 39.36 |
| Total | 96 51.06 | 25 13.30 | 67 35.64 | 188 100.00 |

**Figure 2.13.** *Results from the procedure FREQ: table smoke per race*

```
PROC FREQ data= birthwt;
TABLES smoke*race/CHISQ;
format race ethnicity. smoke tobacco.;
RUN;
```

The CHISQ option gives the result of the chi-square test of independence. The statistic of the chi-square test is equal to 21.7371. The degree of significance of the test is < 0.0001. Tobacco consumption during pregnancy is significantly different depending on race. The consumption is higher in the WHITE group (51.06) than in the BLACK group (13.30) or OTHER (36.64).

Statistics for Table of smoke by race

| Statistic | DF | Value | Prob |
|---|---|---|---|
| Chi-Square | 2 | 21.7371 | <.0001 |
| Likelihood Ratio Chi-Square | 2 | 22.9955 | <.0001 |
| Mantel-Haenszel Chi-Square | 1 | 21.4803 | <.0001 |
| Phi Coefficient | | 0.3400 | |
| Contingency Coefficient | | 0.3219 | |
| Cramer's V | | 0.3400 | |

Sample Size = 188

**Figure 2.14.** *Results from the FREQ procedure: chi-square test*

## 2.2.2. *Scatter plot graph*

Among simple graphical procedures, PROC GPLOT can be used to obtain a representation of a scatter plot graph.

```
PROC GPLOT data= birthwt;
PLOT lwt*bwt;
RUN;
```

The result is displayed below.

The same graph can be obtained with the SGPLOT procedure, more recent and more powerful.

```
PROC SGPLOT data= birthwt;
SCATTER x=bwt y=lwt;
RUN;
```

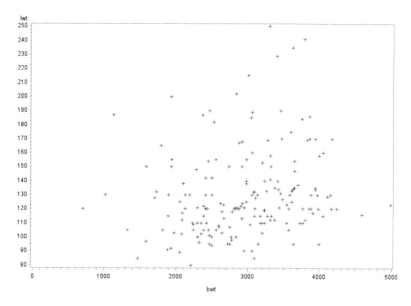

**Figure 2.15.** *Scatter plot graph of variable lwt according to bwt*

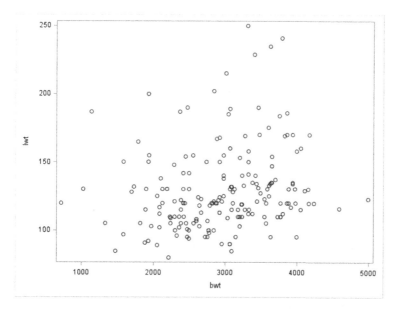

**Figure 2.16.** *Scatter plot graph of variable lwt according to bwt*

The following program allows the regression line of Y in x to be represented on the scatter plot graph.

```
PROC SGPLOT data= birthwt;
SCATTER x=bwt y=lwt;
REG x=bwt y=lwt;
RUN;
```

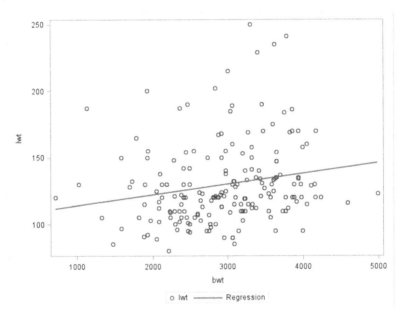

**Figure 2.17.** *Scatter plot graph and regression line of lwt according to bwt*

Note that, only PROC SGPLOT works in the SAS Studio environment, the free version of SAS.

## 2.3. Key points to remember

In this chapter, we have introduced simple procedures of the SAS language for descriptive statistics.

Examples of simple programs are presented including procedures: PROC IMPORT, PROC FREQ, PROC SUMMARY, PROC MEANS, PROC TTEST, PROC GCHART, PROC GPLOT and PROC SGPLOT:

– the PROC FREQ procedure generates simple univariate tabulations of categorical variables;

– the tabulation of simple descriptive statistics for a quantitative variable can be carried out by PROC SUMMARY or PROC MEANS;

– the Student's t-test is carried out by the PROC TTEST procedure. The PAIRED statement is used to perform the test with paired sets;

– the PROC FREQ procedure can also produce contingency tables crossing two or more variables;

– by default, frequencies, percentages, row percentages and column percentages are displayed at the cell level in these contingency tables;

– the ODS GRAPHICS ON statement before each procedure allows, by default, certain graphic outputs illustrating the univariate or bivariate distributions described by these procedures;

– PROC GHART, PROC GPLOT PROC and PROC SGPLOT can be used to generate various histograms and graphs.

## 2.4. Further information

Again, the SAS Base reference manual [SAS 15], provides a detailed description of the capabilities of the procedures for descriptive statistics, introduced in this chapter.

PROC TABULATE [HAW 99] is another powerful procedure useful for descriptive statistics and reporting.

For an in-depth treatment of the analysis of variance with one or two factors and of the analysis of variance for repeated measures, the reader is invited to consult the book by William Dupont [DUP 09].

## 2.5. Applications

1) This example follows Example 3 of Chapter 1.

a) Give the mean and the range (min/max) of the scores of differences per treatment group.

Using PROC SUMMARY summarizes the distribution of the difference scores (mean and range) per treatment group. It should be noted that since a grouping variable is involved, it is necessary in a first stage to sort the data by class (group variable).

```
PROC SORT  DATA=anorexia;  BY group;  RUN;
PROC SUMMARY DATA=anorexia  PRINT  n  mean  min  max  range;
VAR diff; BY group; FORMAT group therapy.; RUN;
```

2) A quantitative variable X takes the following values with a sample of 26 subjects:

```
24.9,25.0,25.0,25.1,25.2,25.2,25.3,25.3,25.3,25.4,25.4,25.4,
25.4,25.5,25.5,25.5,25.5,25.6,25.6,25.6,25.7,25.7,25.8,25.8,
25.9,26.0
```

a) Calculate the mean, the median as well as the mode of X.

b) What is the value of the variance estimated from these data?

c) Assuming that data are grouped into four classes whose bounds are:

```
24.9 to 25.1,  25.2 to 25.4,  25.5 to 25.7,  25.8 to 26.0
```

display the frequency distribution per class in the form of a frequency table.

d) Represent the distribution of X as a histogram, without *a priori* consideration of class intervals.

The capture of raw data will be achieved as in Examples 1 and 2 by means of a DATA step (cards' command)

```
DATA X; INPUT  X; CARDS;
24.9
25.0
25.0
25.1
25.2
25.2
25.3
25.3
25.3
25.4
25.4
```

```
25.4
25.4
25.5
25.5
25.5
25.5
25.6
25.6
25.6
25.7
25.7
25.8
25.8
25.9
26.0;
RUN;
```

The measures of central tendency (mean, median and mode) can be calculated and displayed with PROC SUMMARY.

```
PROC SUMMARY  DATA=X PRINT  mean  median  mode;  VAR X;  RUN;
```

To obtain the variance, we will simply change the calculation options in PROC SUMMARY:

```
PROC SUMMARY  DATA=X PRINT  var;  VAR X; RUN;
```

We could also have added "var" in the PRINT option of the previous PROC SUMMARY.

When re-encoding the variable X into four classes of pre-defined intervals, an auxiliary variable X_classes is created to which the values 1, 2, 3 and 4 are assigned based on the values taken by X.

```
DATA X;  SET  X;
/*  24.9-25.1,  25.2-25.4,  25.5-25.7,  25.8-26.0  */
X_classes=1;
IF  X  ge 25.2  THEN X_classes=2;
IF  X  ge 25.5  THEN X_classes=3;
IF X ge 25.8 THEN X_classes=4;
RUN;
```

The frequency table per class can be directly obtained with a TABLES statement within PROC FREQ:

```
PROC FREQ DATA=X; TABLES X_classes; RUN;
```

To facilitate the reading of the results, we can associate more informative labels to the four classes with PROC FORMAT and associate these with the result table returned by TABLES.

```
PROC FORMAT;
VALUE classes 24.9-25.1="Interval  1"
25.2-25.4="Interval  2"
25.5-25.7="Interval  3"
25.8-26.0="Interval  4";
RUN;
PROC FREQ DATA=X; TABLES X; FORMAT X classes.; RUN;
```

Finally, concerning representations in histograms, SAS has its own algorithms for determining the number of classes that has to be built. Everything can be managed with the PROC GCHART command.

```
PROC GCHART DATA=X; HBAR X; RUN;
```

It is also possible to use a vertically oriented bar chart by specifying the VBAR statement instead of the HBAR (V for vertical, H for horizontal).

```
PROC GCHART DATA=X; VBAR X; RUN;
```

3) The file elderly.dat contains the size of 351 female elderly persons measured in cm, randomly selected from the population during a study on osteoporosis. A few observations are nonetheless missing.

a) How many missing observations can be counted in total?

b) Give a 95% confidence interval for the mean size in this sample, utilizing a normal approximation.

c) Represent the distribution of the sizes observed in the form of a density curve.

To import data stored in a simple text file, the INFILE statement has to be used within a DATA step.

```
DATA elderly;
INFILE      "C:\data\elderly.dat"  dlm="09"X ;
INPUT       size  @@ ;
RUN;
```

It should be noted that the way how missing values are coded is not specified here because the ". " is the format employed, by default, by SAS. Another solution would be to introduce a DATALINE statement after the INPUT statement and to copy and paste the data from the text file. The total number of missing observations can be obtained from PROC SUMMARY by specifying the "nmiss" option.

```
PROC SUMMARY  DATA=elderly  PRINT  nmiss;  VAR size;  RUN;
```

To obtain the mean size and its associated 95% confidence interval, it simply suffice to enter the following command:

```
PROC SUMMARY  DATA=elderly  PRINT  clm  uclm  lclm;
VAR size;
RUN;
```

Finally, to display the distribution of the sizes in the form of a density curve, we use the PROC UNIVARIATE and HISTOGRAM commands with the option /kernel. The degree of smoothing can be controlled by including the option C.

```
PROC UNIVARIATE  DATA=elderly;
VAR size;
HISTOGRAM size / kernel.
RUN;
```

# 3

## Measures of Association, Comparison of Means or Proportions

In this chapter, we are going to present in more detail the statistical procedures seen in the previous chapter: PROC FREQ and PROC TTEST. We will see how they enable us to compare theoretical means or probability of events (theoretical proportions). We will also see how to use the PROC FREQ procedure to estimate odds ratios and other relative risks in a contingency table. Finally, we will address the PROC ANOVA and PROC GLM procedures to carry out multifactor analysis of variance, as well as PROC NPAR1WAY to perform non-parametric tests.

## 3.1. Comparison of two means

### 3.1.1. *Comparison of two means, independent samples*

The PROC TTEST procedure has already been presented in the previous chapter. The results of the following examples have been presented. We recall them, to comment on them as well as on graphics obtained by default.

```
PROC TTEST data= birthwt;
CLASS smoke;
VAR age;
format smoke tobacco.;
RUN;
```

The results of the comparison of variances are given therein (in this case, they do not significantly differ) in addition to the result of the test in both cases: whether these variances be different or not. The Satterthwaite approximation for degrees of freedom is also given. It is advisable to use it when variances significantly differ, with an important *p-value*.

The TTEST Procedure

Variable: age

| smoke | N | Mean | Std Dev | Std Err | Minimum | Maximum |
|---|---|---|---|---|---|---|
| No tobacco consumption during pregnancy | 114 | 23.4649 | 5.4759 | 0.5129 | 14.0000 | 45.0000 |
| Tobacco consumption during pregnancy | 74 | 22.9459 | 5.0474 | 0.5868 | 14.0000 | 35.0000 |
| Diff (1-2) | | 0.5190 | 5.3119 | 0.7930 | | |

| smoke | Method | Mean | 95% CL Mean | | Std Dev | 95% CL Std Dev | |
|---|---|---|---|---|---|---|---|
| No tobacco consumption during pregnancy | | 23.4649 | 22.4488 | 24.4810 | 5.4759 | 4.8456 | 6.2962 |
| Tobacco consumption during pregnancy | | 22.9459 | 21.7766 | 24.1153 | 5.0474 | 4.3449 | 6.0231 |
| Diff (1-2) | Pooled | 0.5190 | -1.0454 | 2.0833 | 5.3119 | 4.8225 | 5.9126 |
| Diff (1-2) | Satterthwaite | 0.5190 | -1.0197 | 2.0577 | | | |

| Method | Variances | DF | t Value | Pr > |t| |
|---|---|---|---|---|
| Pooled | Equal | 186 | 0.65 | 0.5136 |
| Satterthwaite | Unequal | 164.95 | 0.67 | 0.5064 |

| Equality of Variances | | | | |
|---|---|---|---|---|
| Method | Num DF | Den DF | F Value | Pr > F |
| Folded F | 113 | 73 | 1.18 | 0.4562 |

**Figure 3.1.** *Results from the execution of the TTEST procedure*

Another factor to consider in the choice of test to perform is the normality of the distribution. Answers to this question can be found in the graphs. In the previous graph, the histograms of the age variable are represented in each group, as well as the normal densities, with mean the empirical mean and with variance the empirical variance. These two densities are the continuous blue curves. The red dotted curves are non-parametric estimates of the unknown densities in each group of the variable being considered (here, the age).

The box plots or whisker boxes are also represented in each group.

Finally, normality can be better judged with Q-Q plots (for Quantile-Quantile). Under the normality assumption, the points represented are aligned on the first bisector.

### 3.1.2. *Comparison of two means, paired samples*

The syntax of the Student's t-test with paired series is as follows. In the following example, where Lwt and Bwt are the two paired series, the code will be:

```
PROC TTEST data= birthwt;
PAIRED lwt*bwt;
run;
```

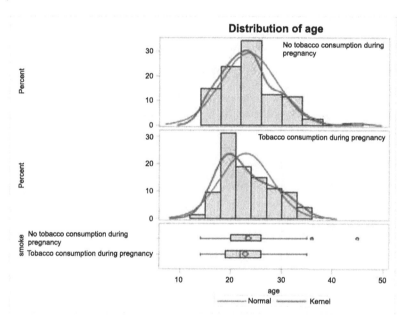

**Figure 3.2.** *Continuation of the results from the execution of the TTEST procedure. For a color version of this figure, see www.iste.co.uk/lalanne/biostatistics3.zip*

The results of the test and the graphics are given in the following.

Note that we can achieve this test by calculating in a DATA step the variable containing the differences between the two paired series, then by making use of PROC UNIVARIATE, which gives by default the Student's t-test for the mean equal to zero.

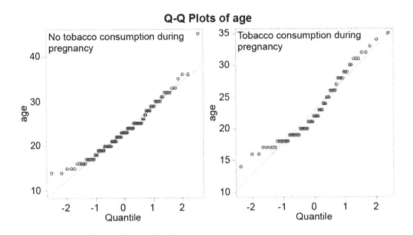

**Figure 3.3.** *Continuation of the results from the execution of the TTEST procedure*

The TTEST Procedure

Difference: lwt - bwt

| N | Mean | Std Dev | Std Err | Minimum | Maximum |
|---|------|---------|---------|---------|---------|
| 188 | -2817.3 | 725.3 | 52.8951 | -4867.0 | -589.0 |

| Mean | 95% CL Mean | | Std Dev | 95% CL Std Dev | |
|------|-------------|---|---------|----------------|---|
| -2817.3 | -2921.6 | -2712.9 | 725.3 | 658.6 | 807.0 |

| DF | t Value | Pr > |t| |
|----|---------|----------|
| 187 | -53.26 | < .0001 |

**Figure 3.4.** *Results from the execution of the TTEST procedure with paired series*

By default, the graphic outputs of PROC TTEST with the PAIRED statement are slightly different from the non-paired case. In addition to the histogram and the Q-Q plot of the differences, the individual profiles are represented (values before and after), as well as the concordance plot for interpreting intra-observation correlation.

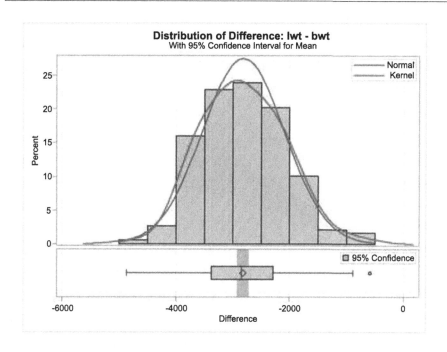

**Figure 3.5.** *Continuation of the results from the execution of the TTEST procedure with paired series. For a color version of this figure, see www.iste.co.uk/lalanne/biostatistics3.zip*

It is possible to perform several PAIRED instructions through a single one, by way of the rule below.

```
The PAIRED statement in the TTEST procedure
-----------------------------------------------------
These PAIRED statements ...
PAIRED A\*B;
PAIRED A\*B C\*D;
PAIRED (A B)\*(C D);
PAIRED (A B)\*(C B);
PAIRED (A1-A2)\*(B1-B2);
PAIRED (A1-A2):(B1-B2);
```

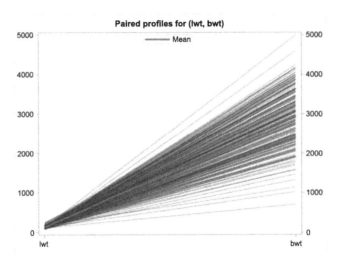

**Figure 3.6.** *Continuation of the results from the execution of the TTEST procedure with paired series. For a color version of this figure, see www.iste.co.uk/lalanne/biostatistics3.zip*

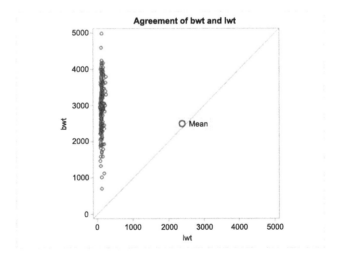

**Figure 3.7.** *Continuation of the results from the execution of the TTEST procedure with paired series*

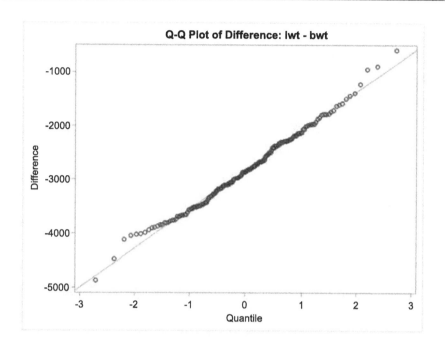

**Figure 3.8.** *Continuation of the results from the execution of
the TTEST procedure with paired series*

### 3.1.3. *Non-parametric Mann-Whitney test*

The PROC NPAR1WAY procedure allows the Wilcoxon test
(Mann-Whitney) to be performed.

```
PROC NPAR1WAY data= birthwt;
CLASS smoke;
VAR age;
format smoke tobacco.;
RUN;
```

The procedure gives, by default, a large number of results and graphics.
We show here only some of these results. The result of the Wilcoxon Mann-
Whitney test is also given. The difference is not significant.

**The NPAR1WAY Procedure**

| Analysis of Variance for Variable age<br>Classified by Variable smoke | | |
| --- | --- | --- |
| smoke | N | Mean |
| No tobacco consumption during pregnancy | 114 | 23.464912 |
| Tobacco consumption during pregnancy | 74 | 22.945946 |

| Source | DF | Sum of Squares | Mean Square | F Value | Pr > F |
| --- | --- | --- | --- | --- | --- |
| Among | 1 | 12.085290 | 12.085290 | 0.4283 | 0.5136 |
| Within | 186 | 5248.143433 | 28.215825 | | |
| Average scores were used for ties. | | | | | |

**Figure 3.9.** *Results from the execution of the NPAR1WAY procedure*

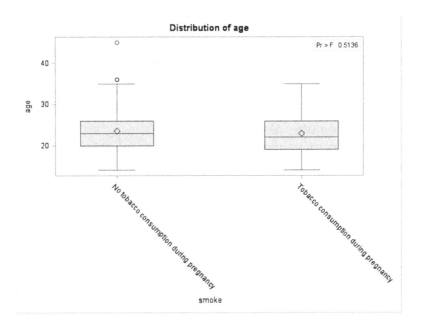

**Figure 3.10.** *Continuation of the results from the execution of the NPAR1WAY procedure*

| Wilcoxon Scores (Rank Sums) for Variable age Classified by Variable smoke | | | | | |
|---|---|---|---|---|---|
| smoke | N | Sum of Scores | Expected Under H0 | Std Dev Under H0 | Mean Score |
| No tobacco consumption during pregnancy | 114 | 11030.50 | 10773.0 | 363.790502 | 96.758772 |
| Tobacco consumption during pregnancy | 74 | 6735.50 | 6993.0 | 363.790502 | 91.020270 |
| Average scores were used for ties. | | | | | |

| Wilcoxon Two-Sample Test | |
|---|---|
| Statistic | 6735.5000 |
| | |
| Normal Approximation | |
| Z | -0.7065 |
| One-Sided Pr < Z | 0.2400 |
| Two-Sided Pr > |Z| | 0.4799 |
| | |
| t Approximation | |
| One-Sided Pr < Z | 0.2404 |
| Two-Sided Pr > |Z| | 0.4808 |
| Z includes a continuity correction of 0.5. | |

**Figure 3.11.** *Continuation of the results from the execution of the NPAR1WAY procedure*

### 3.1.4. *Non-parametric Wilcoxon sign test*

This test is the non-parametric counterpart of the Student's t-test with paired samples. In order to perform it, the difference between the paired series to be compared has to be calculated within a DATA step and then PROC UNIVARIATE has to be employed, which by default and for any submitted variable perfoms the test. The example that follows is an illustration thereof. PROC UNIVARIATE gives the Student's t-test with paired series, as well as two non-parametric tests, including the appropriate Wilcoxon test (test of signed ranks).

```
DATA ; SET birthwt;
diff =lwt-bwt;
PROC UNIVARIATE ;
VAR diff ;
RUN;
```

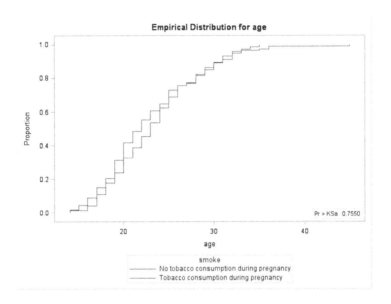

**Figure 3.12.** *Continuation of the results from the execution of the NPAR1WAY procedure. For a color version of this figure, see www.iste.co.uk/lalanne/biostatistics3.zip*

## 3.2. Comparisons of two proportions with independent samples

### 3.2.1. *Chi-square test for the independence between two qualitative variables*

The PROC FREQ procedure is the appropriate procedure as shown in the example that follows:

```
PROC FREQ DATA = birthwt;
TABLES smoke*low/CHISQ ;
FORMAT smoke tobacco. low low.;
run;
```

The test for the comparison of the percentages of baby births with low weight, according to the smoking status is achieved by means of a chi-square test. A number of other equivalent tests are also given by default (likelihood ratio test) and a chi-square test with continuity correction and Fisher's exact test (these last two are to be considered in the presence of low theoretical counts).

The UNIVARIATE Procedure
Variable: diff

| Moments | | | |
|---|---|---|---|
| N | 188 | Sum Weights | 188 |
| Mean | -2817.2926 | Sum Observations | -529651 |
| Std Deviation | 725.260561 | Variance | 526002.882 |
| Skewness | 0.22323756 | Kurtosis | -0.0182821 |
| Uncorrected SS | 1590544357 | Corrected SS | 98362538.9 |
| Coeff Variation | -25.743175 | Std Error Mean | 52.8950628 |

| Basic Statistical Measures | | | |
|---|---|---|---|
| Location | | Variability | |
| Mean | -2817.29 | Std Deviation | 725.26056 |
| Median | -2838.00 | Variance | 526003 |
| Mode | -3839.00 | Range | 4278 |
| | | Interquartile Range | 1093 |

Note: The mode displayed is the smallest of 13 modes with a count of 2.

**Figure 3.13.** *Results from the execution of the UNIVARIATE procedure*

| Tests for Location: Mu0=0 | | | |
|---|---|---|---|
| Test | Statistic | | p Value |
| Student's t | t | -53.2619 | Pr > \|t\| | <.0001 |
| Sign | M | -94 | Pr >= \|M\| | <.0001 |
| Signed Rank | S | -8883 | Pr >= \|S\| | <.0001 |

**Figure 3.14.** *Continuation of the results from the execution
of the UNIVARIATE procedure*

The results in the contingency table that appear in the cells are in this order: observed counts, observed percentage, row percentage and column percentage.

The FREQ Procedure

| Frequency Percent Row Pct Col Pct | Table of smoke by low | | | |
|---|---|---|---|---|
| | | low | | |
| | smoke | Weight greater than 2.5 Kg | Weight less than 2.5 Kg | Total |
| | No tobacco consumption during pregnancy | 85 45.21 74.56 65.89 | 29 15.43 25.44 49.15 | 114 60.64 |
| | Tobacco consumption during pregnancy | 44 23.40 59.46 34.11 | 30 15.96 40.54 50.85 | 74 39.36 |
| | Total | 129 68.62 | 59 31.38 | 188 100.00 |

**Figure 3.15.** *Results from the execution of the FREQ procedure*

It is possible to modify these outputs. Therefore, for example, the following program will print the table containing observed and theoretical counts only.

```
PROC FREQ DATA=birthwt;
TABLES smoke*low/EXPECTED NOCOL NOROW NOPERCENT ;
FORMAT smoke tobacco. low low.;
run;
```

It is also possible to request that several tables be printed at the same time (as well as statistics based on them).

```
Simultaneous printing
```

| Statement | Equivalent statement |
|---|---|
| tables A*(B C) | tables A*B A*C |
| tables (A B)*(C D) | tables A*C B*C A*D B*D |
| tables (A B C)*D | tables A*D B*D C*D |
| tables A - - C | tables A B C |
| tables (A - - C)*D | tables A*D B*D C*D |

**Statistics for Table of smoke by low**

| Statistic | DF | Value | Prob |
|---|---|---|---|
| Chi-Square | 1 | 4.7525 | 0.0293 |
| Likelihood Ratio Chi-Square | 1 | 4.7007 | 0.0302 |
| Continuity Adj. Chi-Square | 1 | 4.0770 | 0.0435 |
| Mantel-Haenszel Chi-Square | 1 | 4.7272 | 0.0297 |
| Phi Coefficient | | 0.1590 | |
| Contingency Coefficient | | 0.1570 | |
| Cramer's V | | 0.1590 | |

| Fisher's Exact Test | |
|---|---|
| Cell (1,1) Frequency (F) | 85 |
| Left-sided Pr <= F | 0.9901 |
| Right-sided Pr >= F | 0.0222 |
| | |
| Table Probability (P) | 0.0123 |
| Two-sided Pr <= P | 0.0366 |

Sample Size = 188

**Figure 3.16.** *Continuation of the results from the
execution of the FREQ procedure (tests of association)*

We can also request that simple histograms be printed (here the observed counts).

```
PROC FREQ DATA=birthwt;
TABLES smoke*low/ PLOTS = freqplot ;
format smoke tobacco. low low.;
run;
```

The FREQ Procedure

| Frequency Expected | Table of smoke by low | | | |
|---|---|---|---|---|
| | | low | | |
| smoke | | Weight greater than 2.5 Kg | Weight less than 2.5 Kg | Total |
| No tobacco consumption during pregnancy | | 85 78.223 | 29 35.777 | 114 |
| Tobacco consumption during pregnancy | | 44 50.777 | 30 23.223 | 74 |
| Total | | 129 | 59 | 188 |

**Figure 3.17.** *Results from the execution of the FREQ procedure (observed and theoretical counts)*

### 3.2.2. *Reading contingency tables based on their counts*

In the following example, the counts of the contingency table are directly read in one data step. The input statement reads the variable Region (numeric variable) and then follows with Eyes and Hair, which are alphabetic variables. Finally, it ends with the variable Count that contains the count of the contingency table cells. The statement LABEL assigns labels to variables that have been read and thereafter, DATALINES, an equivalent statement of CARDS, allows reading the data. Note the symbol @@ after count, which can be used to read all available records before starting a new line. The file created by these command lines will include 27 lines, while the number of statistical units (individuals) is much bigger (equal to the total of Count). The PROC FREQ procedure used with the statement WEIGHT provides a means to find the marginal counts of each variable (eyes and hair), as well as that corresponding to their crossing. The option OUT=FreqCount stores the theoretical counts in a file named Freqcount in the same row as the observed counts.

```
DATA Color;
   INPUT Region Eyes $ Hair $ Count @@;
      LABEL Eyes  ='Eye Color'
Hair  ='Hair Color'
Region='Geographic Region';
      DATALINES;
```

```
1 blue   fair    23  1 blue   red      7  1 blue   medium 24
1 blue   dark    11  1 green  fair    19  1 green  red      7
1 green  medium  18  1 green  dark    14  1 brown  fair    34
1 brown  red      5  1 brown  medium  41  1 brown  dark    40
1 brown  black    3  2 blue   fair    46  2 blue   red     21
2 blue   medium  44  2 blue   dark    40  2 blue   black    6
2 green  fair    50  2 green  red     31  2 green  medium  37
2 green  dark    23  2 brown  fair    56  2 brown  red     42
2 brown  medium  53  2 brown  dark    54  2 brown  black   13
;run;
PROC FREQ DATA=Color;
 TABLES Eyes Hair Eyes*Hair / OUT=FreqCount outexpect sparse;
 WEIGHT Count;
 TITLE 'Eye and Hair Color of European Children';
RUN;
```

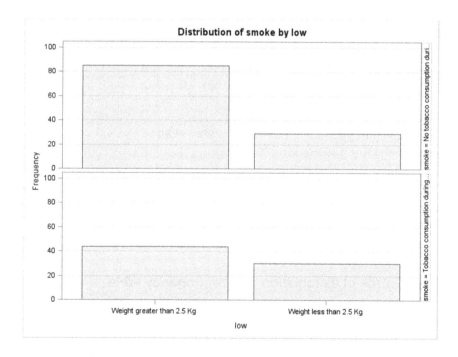

**Figure 3.18.** *Results from the execution of the FREQ procedure (simple histograms for the observed counts)*

**Eye and Hair Color of European Children**

The FREQ Procedure

| Eye Color | | | | |
|---|---|---|---|---|
| Eyes | Frequency | Percent | Cumulative Frequency | Cumulative Percent |
| blue | 222 | 29.13 | 222 | 29.13 |
| brown | 341 | 44.75 | 563 | 73.88 |
| green | 199 | 26.12 | 762 | 100.00 |

| Hair Color | | | | |
|---|---|---|---|---|
| Hair | Frequency | Percent | Cumulative Frequency | Cumulative Percent |
| black | 22 | 2.89 | 22 | 2.89 |
| dark | 182 | 23.88 | 204 | 26.77 |
| fair | 228 | 29.92 | 432 | 56.69 |
| medium | 217 | 28.48 | 649 | 85.17 |
| red | 113 | 14.83 | 762 | 100.00 |

**Figure 3.19.** *Results from the execution of the FREQ procedure: Eyes and Hair frequencies*

```
PROC PRINT DATA=FreqCount noobs;
    TITLE2 'Output Data Set from PROC FREQ';
run;
```

The PROC PRINT procedure prints the file *FreqCount* constructed by PROC FREQ. The *noobs* option suppresses the column in the output that identifies each observation by number.

## 3.3. Measures of association in a contingency table

### 3.3.1. *Odds ratios: Relative risks*

```
PROC FREQ DATA=birthwt;
TABLES smoke*low/RELRISK;
RUN;
```

| Frequency<br>Percent<br>Row Pct<br>Col Pct | Table of Eyes by Hair | | | | | |
|---|---|---|---|---|---|---|
| | | Hair(Hair Color) | | | | |
| Eyes(Eye Color) | black | dark | fair | medium | red | Total |
| blue | 6<br>0.79<br>2.70<br>27.27 | 51<br>6.69<br>22.97<br>28.02 | 69<br>9.06<br>31.08<br>30.26 | 68<br>8.92<br>30.63<br>31.34 | 28<br>3.67<br>12.61<br>24.78 | 222<br>29.13 |
| brown | 16<br>2.10<br>4.69<br>72.73 | 94<br>12.34<br>27.57<br>51.65 | 90<br>11.81<br>26.39<br>39.47 | 94<br>12.34<br>27.57<br>43.32 | 47<br>6.17<br>13.78<br>41.59 | 341<br>44.75 |
| green | 0<br>0.00<br>0.00<br>0.00 | 37<br>4.86<br>18.59<br>20.33 | 69<br>9.06<br>34.67<br>30.26 | 55<br>7.22<br>27.64<br>25.35 | 38<br>4.99<br>19.10<br>33.63 | 199<br>26.12 |
| Total | 22<br>2.89 | 182<br>23.88 | 228<br>29.92 | 217<br>28.48 | 113<br>14.83 | 762<br>100.00 |

**Figure 3.20.** *Results from the execution of the FREQ procedure: table Eyes x Hair*

NOTE.– another procedure, which we will see in Chapter 5, can be used to calculate odds ratios: PROC LOGISTIC.

### 3.3.2. *Comparisons of proportions with non-independent samples: Concordance*

The Kappa option of the FREQ procedure makes it possible to test the equality of proportions from paired samples (square of the difference of the number of discordant pairs proportionally to their total) that SAS calls the McNemar test. The estimate of the concordance Kappa coefficient is then given along with its confidence interval.

```
PROC FREQ DATA=birthwt;
TABLES smoke*low/KAPPA;
RUN;
```

**Eye and Hair Color of European Children
Output Data Set from PROC FREQ**

| Eyes | Hair | COUNT | EXPECTED | PERCENT |
|------|------|-------|----------|---------|
| blue | black | 6 | 6.409 | 0.7874 |
| blue | dark | 51 | 53.024 | 6.6929 |
| blue | fair | 69 | 66.425 | 9.0551 |
| blue | medium | 68 | 63.220 | 8.9239 |
| blue | red | 28 | 32.921 | 3.6745 |
| brown | black | 16 | 9.845 | 2.0997 |
| brown | dark | 94 | 81.446 | 12.3360 |
| brown | fair | 90 | 102.031 | 11.8110 |
| brown | medium | 94 | 97.109 | 12.3360 |
| brown | red | 47 | 50.568 | 6.1680 |
| green | black | 0 | 5.745 | 0.0000 |
| green | dark | 37 | 47.530 | 4.8556 |
| green | fair | 69 | 59.543 | 9.0551 |
| green | medium | 55 | 56.671 | 7.2178 |
| green | red | 38 | 29.510 | 4.9869 |

**Figure 3.21.** *Results from executing the PRINT procedure*

## 3.4. Comparisons of several means

### 3.4.1. *One-way analysis of variance*

One-way analysis of variance makes it possible to simultaneously compare several means. It generalizes the Student's t-test that can achieve it in the case of two means. Two procedures are dedicated to this point: PROC ANOVA and PROC GLM. PROC ANOVA, which is older and more limited, is especially recommended for balanced designs. The two procedures are different when the analysis is multifactorial (more than two unbalanced factors). In addition, PROC GLM allows quantitative factors (analysis of covariance), which PROC ANOVA does not. Qualitative variables (group variables) must be declared with the CLASS statement, which is common to several SAS procedures. The MODEL statement defines the model. In the following example, the quantitative variable lwt is analyzed according to the race factor.

```
PROC ANOVA DATA=birthwt;
CLASS race; MODEL lwt = race;
format race ethnicity.;
RUN;

PROC GLM DATA=birthwt;
CLASS race;
MODEL lwt = race;
format race ethnicity.;
RUN;
```

**Statistics for Table of smoke by low**

| Odds Ratio and Relative Risks | | | |
|---|---|---|---|
| Statistic | Value | 95% Confidence Limits | |
| Odds Ratio | 1.9984 | 1.0676 | 3.7407 |
| Relative Risk (Column 1) | 1.2540 | 1.0098 | 1.5572 |
| Relative Risk (Column 2) | 0.6275 | 0.4130 | 0.9533 |

**Sample Size = 188**

**Figure 3.22.** *Results from the execution of the FREQ procedure with the RELRISK option*

For the example under consideration, the results are almost identical. Once again, it is recommended that the GLM procedure be used.

```
PROC GLM DATA=birthwt;
CLASS race;
MODEL lwt = race;
MEANS race;
format race ethnicity.;
RUN;
```

Statistics for Table of smoke by low

| McNemar's Test | |
| --- | --- |
| Statistic (S) | 3.0822 |
| DF | 1 |
| Pr > S | 0.0792 |

| Simple Kappa Coefficient | |
| --- | --- |
| Kappa | 0.1566 |
| ASE | 0.0725 |
| 95% Lower Conf Limit | 0.0145 |
| 95% Upper Conf Limit | 0.2987 |

Sample Size = 188

**Figure 3.23.** *Results from the execution of the FREQ procedure with the KAPPA option*

The MEANS statement, which must appear after the MODEL statement, makes it possible to print the means and the standard deviations of the response variable (here lwt) for each level of the factor being considered (race, in this case).

```
PROC GLM DATA=birthwt;
CLASS race; MODEL lwt = race;
MEANS race/BON;
FORMAT race ethnicity.;
RUN;
```

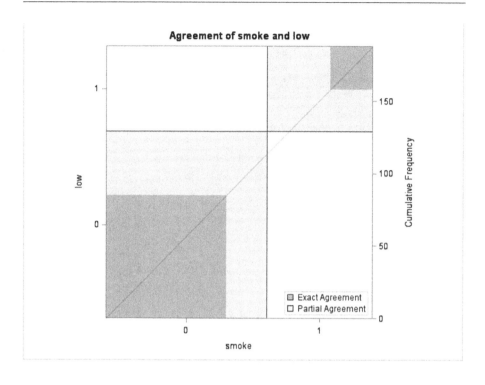

**Figure 3.24.** *Continuation of the results from the execution of the FREQ procedure with the KAPPA option*

The BON (for Bonferroni) option enables multiple comparisons. The confidence intervals of the means that are given are simultaneous. The error risk $\alpha$ (here 0.05) is global. Other methods are possible.

```
PROC GLM DATA=birthwt;
CLASS race; MODEL lwt = race;
MEANS race/HOVTEST;
FORMAT race ethnicity.;
RUN;
```

The option HOVTEST (for Homogeneity of Variances) allows the comparison of the variances of the response variable inside groups. This test is known as the Levene test.

The ANOVA Procedure

Dependent Variable: lwt

| Source | DF | Sum of Squares | Mean Square | F Value | Pr > F |
|---|---|---|---|---|---|
| Model | 2 | 12973.0147 | 6486.5074 | 7.50 | 0.0007 |
| Error | 185 | 160087.7247 | 865.3391 | | |
| Corrected Total | 187 | 173060.7394 | | | |

| R-Square | Coeff Var | Root MSE | lwt Mean |
|---|---|---|---|
| 0.074962 | 22.70903 | 29.41665 | 129.5372 |

| Source | DF | Anova SS | Mean Square | F Value | Pr > F |
|---|---|---|---|---|---|
| race | 2 | 12973.01470 | 6486.50735 | 7.50 | 0.0007 |

**Figure 3.25.** *Results from the execution of the ANOVA procedure*

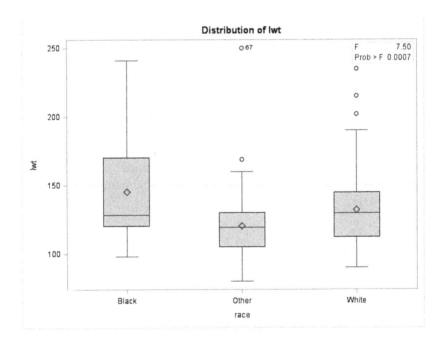

**Figure 3.26.** *Continuation of the results from the execution of the ANOVA procedure*

| Level of race | N | lwt | |
| --- | --- | --- | --- |
| | | Mean | Std Dev |
| Black | 25 | 145.400000 | 39.7879798 |
| Other | 67 | 120.014925 | 25.1302622 |
| White | 96 | 132.052083 | 29.0938119 |

**Figure 3.27.** *Results from the execution of the GLM procedure with the MEANS statement*

Bonferroni (Dunn) t Tests for lwt

Note: This test controls the Type I experimentwise error rate, but it generally has a higher Type II error rate than Tukey's for all pairwise comparisons.

| Alpha | 0.05 |
| --- | --- |
| Error Degrees of Freedom | 185 |
| Error Mean Square | 865.3391 |
| Critical Value of t | 2.41595 |

Comparisons significant at the 0.05 level are indicated by ***.

| race Comparison | Difference Between Means | Simultaneous 95% Confidence Limits | | |
| --- | --- | --- | --- | --- |
| Black - White | 13.348 | -2.610 | 29.306 | |
| Black - Other | 25.385 | 8.729 | 42.041 | *** |
| White - Black | -13.348 | -29.306 | 2.610 | |
| White - Other | 12.037 | 0.724 | 23.351 | *** |
| Other - Black | -25.385 | -42.041 | -8.729 | *** |
| Other - White | -12.037 | -23.351 | -0.724 | *** |

**Figure 3.28.** *Results from the execution of the GLM procedure with the MEANS statement and the BON option*

## 3.4.2. *Two-way analysis of variance*

```
PROC GLM DATA=birthwt;
CLASS race low;
MODEL lwt=race low race*low;
MEANS race low race*low;
FORMAT race ethnicity. low low.;
run;
```

This example deals with two-way analysis of variance with interaction.

The GLM Procedure

| Levene's Test for Homogeneity of lwt Variance ANOVA of Squared Deviations from Group Means | | | | | |
|---|---|---|---|---|---|
| Source | DF | Sum of Squares | Mean Square | F Value | Pr > F |
| race | 2 | 14708448 | 7354224 | 2.25 | 0.1078 |
| Error | 185 | 6.0357E8 | 3262516 | | |

**Figure 3.29.** *Results from the execution of the GLM procedure with the MEANS statement and the HOVTEST option*

The GLM Procedure

Dependent Variable: lwt

| Source | DF | Sum of Squares | Mean Square | F Value | Pr > F |
|---|---|---|---|---|---|
| Model | 5 | 17783.1146 | 3556.6229 | 4.17 | 0.0013 |
| Error | 182 | 155277.6247 | 853.1738 | | |
| Corrected Total | 187 | 173060.7394 | | | |

| R-Square | Coeff Var | Root MSE | lwt Mean |
|---|---|---|---|
| 0.102756 | 22.54884 | 29.20914 | 129.5372 |

| Source | DF | Type I SS | Mean Square | F Value | Pr > F |
|---|---|---|---|---|---|
| race | 2 | 12973.01470 | 6486.50735 | 7.60 | 0.0007 |
| low | 1 | 4723.55696 | 4723.55696 | 5.54 | 0.0197 |
| race*low | 2 | 86.54296 | 43.27148 | 0.05 | 0.9506 |

| Source | DF | Type III SS | Mean Square | F Value | Pr > F |
|---|---|---|---|---|---|
| race | 2 | 12815.20212 | 6407.60106 | 7.51 | 0.0007 |
| low | 1 | 3393.47459 | 3393.47459 | 3.98 | 0.0476 |
| race*low | 2 | 86.54296 | 43.27148 | 0.05 | 0.9506 |

**Figure 3.30.** *Results from the execution of the GLM procedure with two factors*

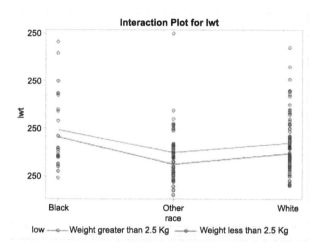

**Figure 3.31.** *Continuation of the results from the execution of the GLM procedure with two factors. For a color version of this figure, see www.iste.co.uk/lalanne/biostatistics3.zip*

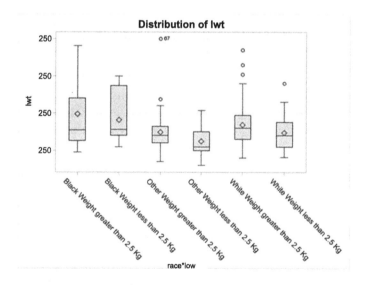

**Figure 3.32.** *Continuation of the results from the execution of the GLM procedure with two factors. For a color version of this figure, see www.iste.co.uk/lalanne/biostatistics3.zip*

## 3.5. Key points to remember

In this chapter, we have continued to utilize, but in a more elaborate way, simple statistical procedures, of which some have already been seen in the previous chapter: PROC FREQ, PROC TTEST, PROC UNIVARIATE, PROC SUMMARY, PROC NPAR1WAY, PROC ANOVA and PROC GLM.

We will start by insisting on this point:

– The ODS GRAPHICS statement on position ON before any procedure enables, by default, certain graphic outputs illustrating the univariate or bivariate distributions (associations) described by these procedures.

– PROC FREQ is the reference procedure for measures and tests of associations between qualitative variables.

– The CHISQ option of the TABLES statement in PROC FREQ produces the chi-square independence test.

– The RELRISK option of the TABLES statement in PROC FREQ generates odds ratios and relative risks.

– The KAPPA option of the TABLES statement in PROC FREQ outputs the concordance Kappa coefficient and provides a mean to test the equality of proportions with paired samples.

– The comparison of two independent samples of a quantitative variable is done by PROC TTEST, which by default generates (if ODS GRAPHICS is in position ON) empirical or estimated histograms, box plots (whiskers plots) and other Q-Q plots.

– The comparison of two paired samples of a quantitative variable is done by PROC TTEST along with the statement PAIRED which will, by default, also produce appropriate graphics.

– The non-parametric Wilcoxon test (Mann-Whitney) is done with PROC NPAR1WAY.

– The non-parametric Wilcoxon sign test is carried out with a DATA step followed by PROC UNIVARIATE.

– The analysis of variance with one or several fixed factors is preferably achieved with PROC GLM.

– In the case where factors are balanced (counts by levels and by crossing equal factors), PROC ANOVA can be used.

– Multiple comparisons of means are made by way of the MEANS statement followed by the BON option.

– The test for the homogeneity of variances is carried out using the MEANS statement followed by the HOVTEST option.

## 3.6. Further information

The help and the examples in the online usage manual of SAS can further assist the reader in finding applications for these procedures which are not described in this book. Similarly to the case of the previous chapter, the SAS Base reference manual [SAS 15], provides a detailed description of the capabilities of PROC FREQ. An important chapter is dedicated to the ODS statement in Ringuede's manual [RIN 14]. Analysis of variance with repeated measures is also carried out by using PROC MIXED, which deals with random factors, not covered in this book.

## 3.7. Applications

1) The birth weight of a sample of 50 children presenting an acute idiopathic respiratory distress syndrome is available. This type of disease can cause death and 27 deaths have been observed among these children. The data are summarized in the table below and are available in the sirds.dat file, where the first 27 observations correspond to the group of children deceased at the time of the study.

Deceased children:

```
1.050 1.175 1.230 1.310 1.500 1.600 1.720 1.750
1.770 2.275 2.500 1.030 1.100 1.185 1.225 1.262
      1.295 1.300 1.550
1.820 1.890 1.940 2.200 2.270 2.440 2.560 2.730
```

Living children:

```
1.130 1.575 1.680 1.760 1.930 2.015 2.090 2.600
2.700 2.950 3.160 3.400 3.640 2.830 1.410 1.715 1.720
      2.040 2.200
 2.400 2.550 2.570 3.005
```

A researcher is interested in whether there exists a difference between the mean weight of children who survived and that of children who died as a consequence of the disease.

a) Perform the Student's t-test. Can we reject the null hypothesis of the lack of differences between the two groups of children?

b) Graphically verify that the application conditions of the test (normality and homogeneity of variances) are verified.

c) What is the 95% confidence interval for the mean difference observed?

Recall that only the numerical values of weights at birth are available in the sirds.dat file, and that we have to build the grouping variable (deceased children vs. alive). One way to proceed consists in performing two DATA steps by including a qualitative variable indicating the status, in addition to the raw data inserted by copy/paste.

```
DATA DCD;
INPUT  weight  @@; death=1; DATALINES;
1.050   1.175   1.230   1.310   1.500   1.600   1.720   1.750   1.770
2.275   2.500   1.030   1.100   1.185   1.225   1.262   1.295   1.300
1.550   1.820   1.890   1.940   2.200   2.270   2.440   2.560   2.730
; RUN;

DATA VIV;
INPUT  weight  @@;
death=0; CARDS;
1.130   1.575   1.680   1.760   1.930   2.015   2.090   2.600   2.700
2.950   3.160   3.400   3.640   2.830   1.410   1.715   1.720   2.040
2.200   2.400   2.550   2.570   3.005
; RUN;
```

DATA DRIA;   SET   DCD  VIV;   RUN;

The Student's t-test is performed using the PROC TTEST command, which by default provides the results under the homoscedasticity or heteroscedasticity assumption, as well as the numerical and graphical descriptive summaries for the distribution of the weight according to the clinical status. The classification variable is introduced after the CLASS statement and the response variable after the VAR statement. By default, the test being exposed is bilateral.

PROC TTEST   DATA=DRIA   PLOTS=all; CLASS   death;
VAR weight; RUN;

2) The sleep quality of 10 patients has been measured before (control) and after treatment by one of the two following hypnotics:  (1) D. hyoscyamine hydrobromide and (2) L. hyoscyamine hydrobromide. The judging criterion chosen by the researchers was the mean gain of sleep (in hours) according to the duration of basic sleep (control) [STU 08]. The data are reported below and are also included in the basic data sets of the software R (data(sleep)).

D. hyoscyamine hydrobromide:
0.7 -1.6 -0.2 -1.2 -0.1   3.4   3.7   0.8   0.0   2.0
L. hyoscyamine hydrobromide:
1.9   0.8   1.1   0.1 -0.1   4.4   5.5   1.6   4.6   3.4

The researchers have concluded that only the second molecule actually had a soporific effect.

a) Estimate the mean sleep time for each of the two molecules, as well as the difference between these two means.

b) Display the distribution of difference scores $(LHH - DHH)$ in the form of a histogram, considering class intervals of half an hour and indicate the mean and the standard deviation of these difference scores.

c) Verify the accuracy of the conclusions with the help of the Student's t-test.

The data from the study on sleep providing a basis for Student's article can be imported in SAS as in the previous exercise, namely by combining the results of two DATA stages. At the same time during the last step, an auxiliary variable will be created for difference scores.

```
DATA DHH;
INPUT   GMSD   @@;
CARDS;
0.7  -1.6  -0.2  -1.2  -0.1  3.4  3.7  0.8  0.0  2.0
; RUN;
DATA LHH;
INPUT   GMSL   @@; CARDS;
1.9  0.8  1.1  0.1  -0.1  4.4  5.5  1.6  4.6  3.4
; RUN;
DATA HH;   MERGE   DHH LHH;   diff_GMS=GMSL-GMSD;   RUN;
```

A numerical summary for the set of numeric variables (GMSD, GMSL and diff_GMS) can be obtained with PROC SUMMARY in the following way:

```
PROC SUMMARY DATA=HH    PRINT  n   mean   var   lclm   uclm;
VAR GMSD   GMSL   diff_GMS;
RUN;
```

Mean gains in sleep time for each molecule can be represented using a bar chart with PROC GCHART.

```
PROC GCHART DATA=hh;
Hbar   diff_gms
 /midpoints=(0   0.5   1   1.5   2   2.5   3   3.5   4   4.5 5 5.5 6);
RUN;
```

Finally, to perform the *t* test for paired data, the PROC TTEST procedure will always be used, but by specifying the option PAIRED, as shown hereafter.

```
PROC TTEST   DATA=hh;
PAIRED   GMSD*GMSL;
RUN;
```

3) In a clinical trial, we tried to evaluate a therapy supposed to reduce the number of symptoms associated with a benign breast disease. A group of 229 women having this disease have been randomly divided into two groups. The first group received routine care, while patients in the second group were following a special therapy (variable B = treatment). One year later, the individuals were assessed and classified into one of the two categories: improvement or no improvement (variable A = response) [SEL 98]. The results are summarized in Table 3.1 for part of the sample.

a) Perform a chi-square test.

b) What are the expected theoretical counts under an hypothesis of independence?

c) Compare the results obtained in (a) with those of the Fisher test.

d) Give a confidence interval for the difference in proportion of improvement between the two groups of patients.

| | Therapy | No therapy | Total |
|---|---|---|---|
| Improvement | 26 | 21 | 47 |
| No improvement | 38 | 44 | 82 |
| Total | 64 | 65 | 129 |

**Table 3.1.** *Scheme and breast disease*

In the case of so-called "grouped" data, or more generally of an arbitrary contingency table, the count table is usually displayed by merely showing the classification variables and the associated counts in clear fashion. In SAS, we can proceed as follows:

```
DATA symptom;
INPUT  therapy improvement count; CARDS;
1  1  26
0  1  21
1  0  38
0  0  44
; RUN;
PROC FORMAT;  VALUE yesno  1="Yes"  0="No";  RUN;
```

Next, we can answer the three questions with the same command, PROC FREQ, remembering that it is required to use the WEIGHT statement to indicate weighting with the counts.

```
PROC FREQ  DATA=symptom  ORDER=data;
TABLES  improvement  *  therapy  / chisq; WEIGHT  count;
FORMAT therapy improvement yesno.; RUN;
```

In the following program, the solution can be obtained with the Riskdiff option. Note the permutation between variables therapy and improvement to get the proper difference in proportion.

```
PROC FREQ  DATA=symptom  ORDER=data;
```

```
TABLES  therapy*improvement  / Riskdiff;
WEIGHT  count;
FORMAT  therapy  improvement  yesno.;
RUN;
```

4) In a study on the estrogen receptor gene, geneticists have directed their interest toward the relationship between the genotype and the age at diagnosis of breast cancer. The genotype was determined from the two alleles of a sequence restriction polymorphism (1.6 and 0.7 kb), that is three groups of subjects: homozygous patients for the 0.7 kb allele (0.7/0.7), homozygous patients for the 1.6 kb allele (1.6/1.6) and heterozygous patients (1.6/0.7). The data have been collected from 59 patients suffering from breast cancer and are available in the file polymorphism.dta (Stata file) [DUP 09]. Mean data are indicated in Table 3.2.

a) Test the null hypothesis according to which the age at diagnosis does not vary with regard to the genotype using the one-way ANOVA. Represent, in graphic form, the distribution of ages for each genotype.

b) The confidence intervals shown in Table 3.2 have been estimated by assuming the homogeneity of variances; in other words, by utilizing the estimate of the common variance, give the value of these confidence intervals without assuming homoscedasticity.

c) Estimate the differences in means corresponding to the set of all possible combinations of the three genotypes, with an estimation of the associated 95 % confidence interval and a parametric test for evaluating the degree of significance (*p-value*) of the observed difference.

d) Graphically represent group means with 95 % confidence intervals.

| | Genotype | | | Total |
|---|---|---|---|---|
| | 1.6/1.6 | 1.6/0.7 | 0.7/0.7 | |
| Number of patients | 14 | 29 | 16 | 59 |
| *Age at diagnosis* | | | | |
| Mean | 64.64 | 64.38 | 50.38 | 60.64 |
| Standard deviation | 11.18 | 13.26 | 10.64 | 13.49 |
| IC 95 % | (58.1–71.1) | (59.9–68.9) | (44.3–56.5) | |

**Table 3.2.** *Estrogen receptor gene polymorphism*

Data in Stata format can be imported into SAS by using PROC IMPORT and indicating the type of data source, here DBMS=STATA.

```
PROC IMPORT  OUT= WORK.polymorphism
DATAFILE=  "C:\data\polymorphism.dta" DBMS=STATA  REPLACE;
RUN;
```

The PROC GLM command is used for linear models (ANOVA, simple or multiple linear regression, ANCOVA). In the case of the one-way ANOVA, the classification factor is indicated after the CLASS statement and the ANOVA model after the MODEL statement. The one-way ANOVA model will therefore be written as $y = g$, where $y$ denotes the response variable and $g$ the study factor.

```
PROC GLM  DATA=polymorphism; CLASS  genotype;
MODEL  age=genotype; RUN;
```

The means of groups with their 95% confidence intervals can be obtained using PROC SUMMARY. The conditioning variable is specified after the BY statement.

```
PROC SORT  DATA=polymorphism; BY genotype;  RUN;
PROC SUMMARY  DATA=polymorphism PRINT n  mean stddev ucl lcl;
VAR age;
BY genotype;
RUN;
```

Regarding multiple comparisons, we always use PROC GLM, by adding this time the option MEANS genotype/BON CLDIFF that provides the tests corrected by the Bonferroni method (also called Dunn tests in SAS).

```
PROC GLM  DATA=polymorphism; CLASS genotype;
MODEL  age=genotype;
MEANS  genotype  / BON CLDIFF;
RUN;
```

The CLDIFF option makes it possible to represent results related to the differences in means in the form of confidence intervals, whereas CLM does the same for group means (simultaneous confidence intervals).

```
PROC GLM  DATA=polymorphism  PLOT=MEANPLOT(CLBAND);
CLASS  genotype;
MODEL  age=genotype;
```

```
MEANS  genotype  / BON CLM;
RUN;
```

The PROC GLM command also provides graphic outputs, including a representation of the distribution of age according to the genotype in the form of box plots. It is also possible to represent the age distribution according to the genotype as a bar chart.

```
PROC SORT  DATA=polymorphism;  BY genotype;  RUN;
PROC GCHART DATA=polymorphism;  Vbar  age;  BY genotype;  RUN;
```

5) An obstetric service is interested in the weight of full term and 1-month-old infants [PEA 05]. For this sample of 550 babies, information concerning parity (number of brothers and sisters) is also available; however, we know that there is no twinhood relationship among children who have brothers and sisters. The purpose of the study is to determine whether parity (four classes) influences the weight of 1-month-old newborn babies. The data are summarized in Table 3.3 and they are available in an SPSS file named weights.sav.

a) Verify the data represented in Table 3.3.

b) Perform the one-way analysis of variance. Conclude with the overall significance and indicate the proportion of variance explained by the model.

c) Display the weight distribution according to parity. Conduct a homogeneity test of variances (search the online help for the Levene test).

d) The last two categories (2 and $\geq$ 3) were regrouped. Redo the analysis and compare it to the results obtained in (b).

e) Perform a linear trend test (with ANOVA) with the data recoded into three levels for parity.

The file containing the data, weights.sav, has been exported from R in Stata format using the following R commands:

```
library(foreign)
weights  <-  read.spss("weights.sav",  to.data.frame=TRUE)
write.dta(weights,  file="weights.dta")
```

Therefore, it is possible to import it with PROC IMPORT, by instructing the type of data source (DBMS=STATA), as it has been done for Example 4.

```
PROC IMPORT   OUT= WORK.weight
DATAFILE=  "C:\data\weights.dta" DBMS=STATA  REPLACE;
RUN;
```

| | Number of brothers and sisters | | | | Total |
|---|---|---|---|---|---|
| | 0 | 1 | 2 | $\geq 3$ | |
| *Sample* | | | | | |
| Count | 180 | 192 | 116 | 62 | 550 |
| Frequency | 32.7 | 34.9 | 21.1 | 11.3 | 100.0 |
| *Weight (kg)* | | | | | |
| Mean | 4.26 | 4.39 | 21.1 | 11.3 | |
| Standard deviation | 0.62 | 0.59 | 0.61 | 0.54 | |
| (Min-Max) | (2.92–5.75) | (3.17–6.33) | (3.09–6.49) | (3.20–5.48) | |

**Table 3.3.** *Newborns' weight*

The counts and relative frequencies table for the parity variable is obtained from the PROC FREQ command by specifying the TABLES statement for the variable of interest:

```
PROC FREQ  DATA=weight;  TABLES  parity;  RUN;
```

The means and standard deviations of the weight according to the number of siblings are thus obtained:

```
PROC SORT  DATA=weight;  BY parity;  RUN;
PROC SUMMARY DATA=weight PRINT n mean stddev min max/class parity;
VAR weight;
RUN;
```

Single-factor ANOVA is performed as in previous exercises, for example by means of the PROC GLM command and by providing the response variable as well as the qualitative variable defining the groups to be compared.

```
PROC GLM  DATA=weight; CLASS  parity;
MODEL  weight=parity;
RUN;
```

The weight distribution is displayed according to parity and the test for homogeneity of variances is thus carried out:

```
PROC GLM  DATA=weight; CLASS  parity;
MODEL  weight=parity;
MEANS parity/hovtest;
RUN;
```

By default, the utilized test is the Levene test. To obtain the Bartlett test (another test for homogeneity of variances), it suffices to change the option in the following manner: HOVTEST=BARTLETT.

To re-encode the parity variable into three classes, here follows a possible solution in SAS:

```
DATA weight1; SET weight; Newparity=parity+1-1; RUN;
PROC FREQ DATA=weight1; TABLES newparity; RUN;
DATA weight2; SET weight1; Nparity=newparity;
IF newparity GE 3 THEN Nparity=3;
RUN;
PROC FREQ DATA=weight2; TABLES Nparity; RUN;
```

The analysis of variance model can be estimated again from these new data (weight2) with PROC GLM.

```
PROC GLM DATA=weight2; CLASS Nparity;
MODEL weight=Nparity; MEANS Nparity / HOVTEST;
RUN;
```

A test for the linear trend with the data re-coded on three levels for parity will be achieved by removing the CLASS directive. The variable Nparity will be considered as quantitative, and levels 1, 2 and 3 are ordered. This is a simple linear regression test. The MEANS directive is no longer useful.

```
PROC GLM DATA=weight2;
MODEL weight=Nparity;
RUN;
```

6) The ToothGrowth data set available in R contains data from a study on the length of odontoblasts (variable len) among 10 Guinea pigs after administration of different doses of vitamin C (0.5, 1 or 2 mg, variable dose) in the form of ascorbic acid or orange juice (variable supp) [BLI 52].

a) Verify the distribution of counts according to the different treatment conditions (cross-tabulating the modalities of the two factors, supp and dose) of this experimental design.

b) Calculate the mean and the standard deviation of each treatment.

c) Build an analysis of variance table for the full model with interaction between the two factors.

d) Draw an interaction chart representing the mean values of the response variable according to the levels of the two factors.

The verification of the counts can be done using proc freq.

```
proc freq data=tooth;tables supp*dose/nopercent norow nocol;run;
```

The calculation of the mean and standard deviation of each treatment can be done using:

```
proc summary data=tooth print mean stddev;var len;run;
```

The analysis of variance table for the full model with interaction between the two factors can be obtained by using the program below. The interaction chart is given by default.

```
Proc GLM data=tooth;
class supp dose;
model len=supp dose supp*dose;
run;
```

# 4

# Correlation, Linear Regression

In this chapter, we are essentially going to present two new statistical procedures: PROC CORR and PROC REG.

PROC CORR provides a means to measure the association between two quantitative variables and generates measures of linear association (PEARSON's linear correlation coefficient). It also allows the computation of Spearman's and Kendall's rank correlation coefficients.

The PROC REG procedure is designed for linear regression. Linear regression is a model that assumes that the conditional means of a response variable Y with respect to a variable X are aligned. The regression coefficient can be interpreted as the slope of a line, known as the regression line of Y on X, describing the evolution of Y according to X.

## 4.1. Linear correlation

### 4.1.1. *Pearson's correlation coefficient*

```
PROC CORR data= birthwt;
VAR age lwt bwt;
RUN;
```

The PROC CORR procedure produces simple statistics (counts, means, sum, standard deviation, minimum, maximum) of each variable listed in the VAR statement by default, as well as the PEARSON correlation coefficient and the degree of significance of the test $\rho = 0$. By default, scatter plot graphs are also generated in a matrix form. The HISTOGRAM option below outputs histograms for each of the listed variables.

```
PROC CORR data= birthwt PLOTS=MATRIX (NVAR=ALL HISTOGRAM);
VAR age lwt bwt;
RUN;
```

The CORR Procedure

3 Variables:  age lwt bwt

| | | Simple Statistics | | | | |
|---|---|---|---|---|---|---|
| Variable | N | Mean | Std Dev | Sum | Minimum | Maximum |
| age | 188 | 23.26064 | 5.30373 | 4373 | 14.00000 | 45.00000 |
| lwt | 188 | 129.53723 | 30.42135 | 24353 | 80.00000 | 250.00000 |
| bwt | 188 | 2947 | 730.50775 | 554004 | 709.00000 | 4990 |

| Pearson Correlation Coefficients, N = 188 Prob > \|r\| under H0: Rho=0 | | | |
|---|---|---|---|
| | age | lwt | bwt |
| age | 1.00000 | 0.18917 0.0093 | 0.08807 0.2294 |
| lwt | 0.18917 0.0093 | 1.00000 | 0.19269 0.0081 |
| bwt | 0.08807 0.2294 | 0.19269 0.0081 | 1.00000 |

**Figure 4.1.** *Results from the execution of the CORR procedure*

### 4.1.2. *Other correlation coefficients*

The options SPEARMAN, KENDALL and ALPHA enable the respective calculations of the SPEARMAN non-parametric correlation coefficient (rank correlation coefficient), Kendall's tau coefficient and Cronbach's alpha coefficient.

```
PROC CORR data= birthwt SPEARMAN KENDALL ALPHA ;
VAR age lwt bwt;
RUN;
```

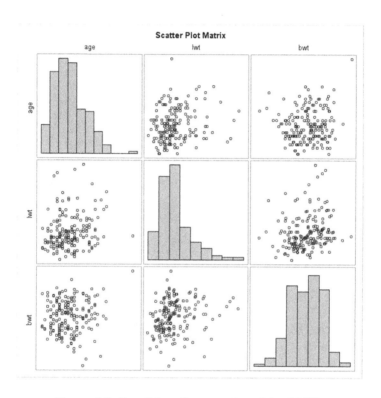

**Figure 4.2.** *Result from the execution of the CORR procedure with the option PLOTS*

### 4.1.3. *Fisher transformation*

The FISHER option makes it possible to test the hypothesis $\rho = \rho_0$, where $\rho_0$ is a fixed given value. In the example below, the chosen value to test is $0.3$.

```
PROC CORR data= birthwt FISHER (RHO0=0.3);
VAR lwt bwt;
RUN;
```

| Cronbach Coefficient Alpha | |
|---|---|
| Variables | Alpha |
| Raw | 0.025669 |
| Standardized | 0.357829 |

| Cronbach Coefficient Alpha with Deleted Variable | | | | |
|---|---|---|---|---|
| | Raw Variables | | Standardized Variables | |
| Deleted Variable | Correlation with Total | Alpha | Correlation with Total | Alpha |
| age | 0.095110 | 0.031536 | 0.179510 | 0.323113 |
| lwt | 0.193931 | 0.002554 | 0.258856 | 0.161890 |
| bwt | 0.198689 | 0.120329 | 0.182054 | 0.318158 |

**Figure 4.3.** *Result from the execution of the CORR procedure with the SPEARMAN KENDALL ALPHA option*

## 4.2. Linear regression

### 4.2.1. *Simple linear regression*

```
PROC REG data= birthwt ;
MODEL lwt= bwt;
RUN;
```

### 4.2.2. *Test for the linearity of the model*

```
PROC REG data= birthwt ;
MODEL lwt= bwt/ LACKFIT;
RUN;
```

The option LACKFIT of the MODEL statement allows the linearity of the model to be tested.

### 4.2.3. *Analysis of covariance*

```
PROC GLM data= birthwt ;
CLASS low;
MODEL lwt= bwt low bwt*low;
RUN;
```

| Spearman Correlation Coefficients, N = 188 Prob > \|r\| under H0: Rho=0 | | | |
| --- | --- | --- | --- |
| | age | lwt | bwt |
| age | 1.00000 | 0.19484 0.0074 | 0.05814 0.4281 |
| lwt | 0.19484 0.0074 | 1.00000 | 0.25617 0.0004 |
| bwt | 0.05814 0.4281 | 0.25617 0.0004 | 1.00000 |

| Kendall Tau b Correlation Coefficients, N = 188 Prob > \|tau\| under H0: Tau=0 | | | |
| --- | --- | --- | --- |
| | age | lwt | bwt |
| age | 1.00000 | 0.13784 0.0068 | 0.03792 0.4522 |
| lwt | 0.13784 0.0068 | 1.00000 | 0.17477 0.0004 |
| bwt | 0.03792 0.4522 | 0.17477 0.0004 | 1.00000 |

**Figure 4.4.** *Continuation of the result from the execution of the CORR procedure with the SPEARMAN KENDALL ALPHA option*

| Pearson Correlation Statistics (Fisher's z Transformation) | | | | | | | | | | |
| --- | --- | --- | --- | --- | --- | --- | --- | --- | --- | --- |
| | | | | | | | | | H0:Rho=Rho0 | |
| Variable | With Variable | N | Sample Correlation | Fisher's z | Bias Adjustment | Correlation Estimate | 95% Confidence Limits | | Rho0 | p Value |
| lwt | bwt | 188 | 0.19269 | 0.19513 | 0.0005152 | 0.19219 | 0.050468 | 0.326325 | 0.30000 | 0.1172 |

**Figure 4.5.** *Continuation of the result from the execution of the CORR procedure with the FISHER option*

PROC GLM can model the mean of a quantitative response variable (here lwt) linearly according to quantitative variables (here bwt) and factors (here low) and their interaction.

The REG Procedure
Model: MODEL1
Dependent Variable: lwt

| Number of Observations Read | 188 |
|---|---|
| Number of Observations Used | 188 |

| Analysis of Variance | | | | | |
|---|---|---|---|---|---|
| Source | DF | Sum of Squares | Mean Square | F Value | Pr > F |
| Model | 1 | 6425.40740 | 6425.40740 | 7.17 | 0.0081 |
| Error | 186 | 166635 | 895.88888 | | |
| Corrected Total | 187 | 173061 | | | |

| Root MSE | 29.93140 | R-Square | 0.0371 |
|---|---|---|---|
| Dependent Mean | 129.53723 | Adj R-Sq | 0.0320 |
| Coeff Var | 23.10641 | | |

| Parameter Estimates | | | | | |
|---|---|---|---|---|---|
| Variable | DF | Parameter Estimate | Standard Error | t Value | Pr > \|t\| |
| Intercept | 1 | 105.89112 | 9.09536 | 11.64 | <.0001 |
| bwt | 1 | 0.00802 | 0.00300 | 2.68 | 0.0081 |

**Figure 4.6.** *Results from the execution of the REG procedure*

The lack of interaction means that the slopes of the lines (lwt with respect to bwt) at each level defined by the factor are equal (the lines are parallel).

In this example, we conclude that there is a marginal positive linear evolution of lwt with respect to bwt. This evolution is non-significant at each level of low. In addition, the means of lwt do not differ according to the level of low.

```
PROC GLM data= birthwt ;
CLASS low;
MODEL lwt= bwt low ;
RUN;
```

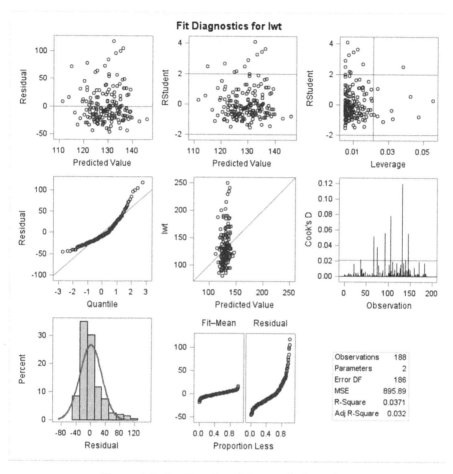

**Figure 4.7.** *Continuation of the results from the execution of the REG procedure*

### 4.2.4. *Multiple linear regression*

Multiple linear regression analyses can be performed with PROC REG or with PROC GLM.

```
PROC REG data= birthwt ;
MODEL lwt= bwt age;
RUN;
PROC GLM data= birthwt ;
MODEL lwt- bwt age;
RUN;
```

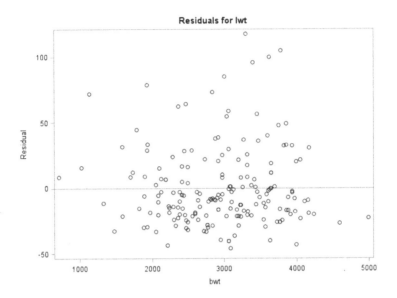

**Figure 4.8.** *Continuation of the results from the execution of the REG procedure*

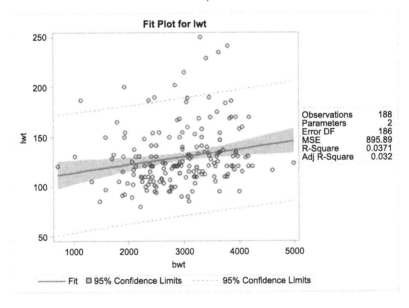

**Figure 4.9.** *Continuation of the results from the execution of the REG procedure*

| Analysis of Variance | | | | | |
|---|---|---|---|---|---|
| Source | DF | Sum of Squares | Mean Square | F Value | Pr > F |
| Model | 1 | 6425.40740 | 6425.40740 | 7.17 | 0.0081 |
| Error | 186 | 166635 | 895.88888 | | |
| Lack of Fit | 128 | 124538 | 972.95663 | 1.34 | 0.1053 |
| Pure Error | 58 | 42097 | 725.80833 | | |
| Corrected Total | 187 | 173061 | | | |

**Figure 4.10.** *Result from the execution of the REG procedure with the option LACKFIT*

The GLM Procedure

Dependent Variable: lwt

| Source | DF | Sum of Squares | Mean Square | F Value | Pr > F |
|---|---|---|---|---|---|
| Model | 3 | 8808.5990 | 2936.1997 | 3.29 | 0.0219 |
| Error | 184 | 164252.1403 | 892.6747 | | |
| Corrected Total | 187 | 173060.7394 | | | |

| R-Square | Coeff Var | Root MSE | lwt Mean |
|---|---|---|---|
| 0.050899 | 23.06492 | 29.87766 | 129.5372 |

| Source | DF | Type I SS | Mean Square | F Value | Pr > F |
|---|---|---|---|---|---|
| bwt | 1 | 6425.407398 | 6425.407398 | 7.20 | 0.0080 |
| low | 1 | 77.833534 | 77.833534 | 0.09 | 0.7681 |
| bwt*low | 1 | 2305.358095 | 2305.358095 | 2.58 | 0.1098 |

| Source | DF | Type III SS | Mean Square | F Value | Pr > F |
|---|---|---|---|---|---|
| bwt | 1 | 109.451592 | 109.451592 | 0.12 | 0.7266 |
| low | 1 | 1925.643436 | 1925.643436 | 2.16 | 0.1436 |
| bwt*low | 1 | 2305.358095 | 2305.358095 | 2.58 | 0.1098 |

**Figure 4.11.** *Results from the execution of the GLM procedure*

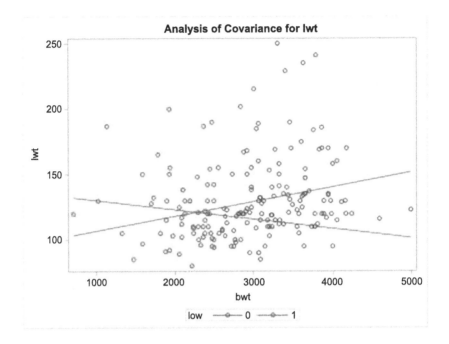

**Figure 4.12.** *Continuation of the results from the execution of the GLM procedure. For a color version of this figure, see www.iste.co.uk/lalanne/biostatistics3.zip*

With the CLASS statement, PROC GLM enables the use of qualitative variables X, which PROC REG does not. PROG REG is more suited to multiple linear regression and it has a powerful option for selecting models. The SELECTION option which can appear after the MODEL statement achieves automatic step-by-step searches for the model that best fits the data, according to several methods: forward selection (FORWARD), backward elimination (BACKWARD), etc. It is also possible to force the inclusion of certain variables in all the models investigated by the SELECTION option, with the INCLUDE option which can appear in the MODEL statement.

### 4.2.5. *Simple non-parametric regressions with the SGPLOT procedure*

```
PROC SGPLOT data=birthwt;
  SCATTER x=lwt y=bwt;
```

```
PBSPLINE x=lwt y=bwt;
LOESS x=lwt y=bwt;
REG x=lwt y=bwt;
RUN;
```

### The GLM Procedure

#### Dependent Variable: lwt

| Source | DF | Sum of Squares | Mean Square | F Value | Pr > F |
|---|---|---|---|---|---|
| Model | 2 | 6503.2409 | 3251.6205 | 3.61 | 0.0289 |
| Error | 185 | 166557.4984 | 900.3108 | | |
| Corrected Total | 187 | 173060.7394 | | | |

| R-Square | Coeff Var | Root MSE | lwt Mean |
|---|---|---|---|
| 0.037578 | 23.16336 | 30.00518 | 129.5372 |

| Source | DF | Type I SS | Mean Square | F Value | Pr > F |
|---|---|---|---|---|---|
| bwt | 1 | 6425.407398 | 6425.407398 | 7.14 | 0.0082 |
| low | 1 | 77.833534 | 77.833534 | 0.09 | 0.7691 |

| Source | DF | Type III SS | Mean Square | F Value | Pr > F |
|---|---|---|---|---|---|
| bwt | 1 | 1792.641631 | 1792.641631 | 1.99 | 0.1599 |
| low | 1 | 77.833534 | 77.833534 | 0.09 | 0.7691 |

**Figure 4.13.** *Results from the execution of the GLM procedure, interactionless model*

## 4.3. Key points to remember

In this chapter, we have introduced two procedures: PROC CORR and PROC REG, as well as demonstrated new possibilities for PROC GLM and PROC SGPLOT:

– PROC CORR produces the Pearson linear correlation coefficient, those of Spearman's and Kendall's ranks;

– PROC CORR also outputs Cronbach's $\alpha$ coefficient;

– PROC REG can generate the linear regression of a univariate response variable Y in one or several explanatory variables X;

– PROC GLM enables the manipulation of categorical explanatory variables, thus to produce a standard analysis of covariance;

– It is possible to generate non-parametric simple regressions with PROC SGPLOT.

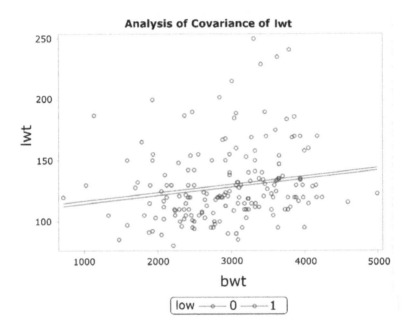

**Figure 4.14.** *Results from the execution of the GLM procedure, interactionless model. For a color version of this figure, see www.iste.co.uk/lalanne/biostatistics3.zip*

## 4.4. Further information

Again, the help and the examples in the online usage manual of SAS can further assist the reader in figuring possibilities for procedures not described in this book.

Linear regression models are also carried out by using PROC MIXED, which deals with random factors and random coefficients, not covered in this book.

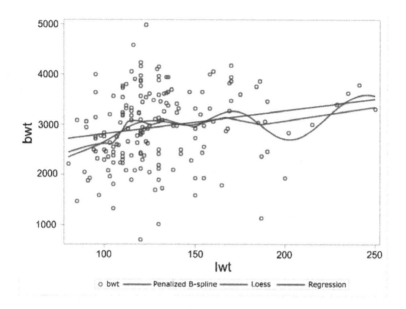

**Figure 4.15.** *Results from the execution of the SGPLOT procedure. For a color version of this figure, see www.iste.co.uk/lalanne/biostatistics3.zip*

## 4.5. Applications

1) A study focused on a measure of malnutrition among 25 patients aged from 7 to 23 years old and suffering from cystic fibrosis. For these patients, varied information was available relatively to anthropometric characteristics (size, weight, etc.) and to lung function [EVE 01]. The data are available in the file cystic.dat.

a) Calculate the linear correlation coefficient between the variables PEmax and Weight, as well as its 95 % confidence interval.

b) Display the set of numerical data in the form of scatter plots, namely 45 plots arranged in the form of a "dispersion matrix".

c) Calculate the set of Pearson's and Spearman's correlations between numeric variables.

d) Calculate the correlation between PEmax and Weight by controlling the age (Age) (partial correlation). Graphically represent the covariance between PEmax and Weight by highlighting the two most extreme terciles for the variable age.

The tabulated data are available in a text file that can be imported by means of a DATA step with an INFILE statement and by specifying that the first line of the file contains the name of the variables.

```
DATA cystic;
INFILE    "C:\data\cystic.dat"  firstobs=2;
INPUT Sub Age Sex Height Weight BMP FEV RV FRC TLC PEmax;RUN;
```

To calculate the value of the BravaisPearson correlation coefficient, it is necessary to use PROC CORR and to instruct the variables of interest as shown hereafter. SAS automatically outputs a graphical output if a statement ODS GRAPHICS ON is specified prior to the launch of the CORR procedure.

```
PROC CORR DATA=cystic  fisher;  VAR PEmax  Weight;    RUN;
```

The "fisher" option tells SAS that the calculations must use an inverse Fisher transformation, which in turn provides an estimate of the 95% confidence interval for the correlation coefficient between the variables PEmax and Weight. The same procedure will be used for the hypothesis test $Ho : \rho = 0$, with for example, as a specific alternative $Ha : \rho = 0.3$. We will simply add an option $RHO0 = 0.3$.

```
PROC CORR DATA=cystic fisher (RHO0=0.3); VAR PEmax Weight; RUN;
```

To display the totality of dispersion diagrams, PROC CORR must be employed by specifying the graphic option (PLOTS=matrix). The list of the study variables is indicated after VAR.

```
PROC CORR DATA=cystic  PLOTS=matrix  (NVAR=all  histogram);
VAR sex Age Height Weight BMP  FEV  RV FRC  TLC PEmax; RUN;
```

As can be seen, SAS provides the value of the correlation coefficient for each pair of variables, as well as the degrec of significance for the associated test for the nullity of the correlation. For Spearman's correlations, the option SPEARMAN has to be added:

```
PROC CORR DATA=cystic  SPEARMAN  KENDALL;
VAR sex Age Height Weight BMP  FEV  RV FRC  TLC PEmax; RUN;
```

The partial correlation between variables PEmax and Weight taking Age into account is obtained by specifying the Partial option followed by the name of the control variable. The PLOTS option makes it possible to graphically represent the covariance between variables PEmax and Weight highlighting the two most extreme terciles for the variable Age.

```
PROC CORR DATA=cystic  PLOTS=scatter(alpha=.33  .66);
VAR PEmax  Weight; PARTIAL  age;  RUN;
```

2) In the Framingham study, data are available on systolic blood pressure (sbp) and on body mass index (bmi) of 2,047 men and 2,643 women [DUP 09]. We are mainly interested in the relationship between these two variables (after logarithmic transformation) in men and in women separately. The data are available in the file Framingham.csv.

a) Graphically represent variations between blood pressure (sbp) and body mass index (bmi) in men and in women.

b) Do linear correlation coefficients estimated for men and women show a significant difference, at the level $\alpha = 5\%$?

c) Estimate the parameters of the linear regression model considering blood pressure as the response variable and BMI as the explanatory variable, for these two subsamples. Give a 95% confidence interval for the estimate of the respective slopes.

Since data are available in "conventional" CSV format (using comma as the field separator), they can be imported very simply by means of PROC IMPORT.

```
PROC IMPORT  OUT= WORK.FRAMINGHAM
DATAFILE=  "C:\data\Framingham.csv" DBMS=CSV  REPLACE;
GETNAMES=YES;
DATAROW=2;
RUN;
```

To graphically represent blood pressure variations (sbp) according to the body mass index (bmi), a scatter plot can be used with PROC SGPLOT.

```
PROC SGPLOT  DATA=framingham;
SCATTER x=bmi  y=sbp / group=sex; REG  x=bmi  y=sbp ;
RUN;
```

The analyses presented below are based on an analysis of covariance approach, in which several models are considered (model with interaction between sex and body mass index, model without interaction). First, the two variables "bmi" and "sbp" will be transformed using a logarithmic transformation within a DATA step.

DATA framingham; SET framingham; logBMI= Log(bmi);
Logsbp = Log(sbp); RUN;

The interaction model is expressed as follows in PROC GLM: sbp=bmi sex bmi*sex, the term to the left of the = sign corresponds to the response variable and the variables to the right are the explanatory variables of the model, the interaction between two variables being symbolized by the symbol *.

PROC GLM DATA=framingham; CLASS sex;
MODEL Logsbp=logBMI sex logBMI*sex; RUN;

The model without interaction is specified on the same principle, by omitting the bmi*sex term.

The regression model stratified by sex is not a major problem and, unlike R, it is not necessary to calculate confidence intervals for slopes with a separate command because these are directly provided in the results table returned by SAS.

PROC SORT  DATA=framingham;  BY sex;  RUN;
PROC REG DATA=framingham;
MODEL  Logsbp=logBMI/clb;  BY sex;  RUN;

PROC GLM DATA=framingham; CLASS sex;
MODEL  Logsbp=logBMI sex logBMI*sex; RUN;

3) The data available in the file quetelet.csv provide information on systolic blood pressure (PAS), the Quetelet index (QTT), age (AGE) and tobacco consumption (TAB=1 if smoking, 0 otherwise) for a sample of 32 men older than 40.

a) Indicate the value of the linear correlation coefficient between systolic blood pressure and the Quetelet index, with a 90% confidence interval.

b) Give the estimates of the parameters of the linear regression line of blood pressure on the Quetelet index.

c) Test whether the slope of the regression line is different from 0 (at the 5 % threshold).

d) Graphically represent blood pressure variations according to the Quetelet index, distinctly showing smokers and non-smokers with different symbols or colors, and draw the regression line whose parameters have been estimated in (b).

e) Repeat analysis (bc) by restricting the sample to smokers.

There are no major difficulties when loading the data because they have been exported from Excel and are in CSV format. The type of field delimiter will, however, have to be specified during the DATA step, which is here a semicolon.

```
PROC IMPORT  OUT= WORK.QUETELET
DATAFILE=  "C:\data\quetelet.txt" DBMS=DLM  REPLACE;
DELIMITER='3B'x; GETNAMES=YES;
DATAROW=2;
RUN;
```

The linear correlation coefficient between the variables "pas" and "qtt" can be estimated, as well as its confidence interval, with the PROC CORR procedure discussed in the previous exercise. The level of confidence can be specified with the option alpha=0.10.

```
PROC CORR DATA=quetelet fisher  (alpha=0.10); VAR PAS QTT;
RUN;
```

The PROC REG command makes it possible to perform a linear regression for a response variable (placed first in the list of variables) and one or more explanatory variables. It is used as follows:

```
PROC REG DATA=quetelet; MODEL pas=qtt; RUN;
```

By default, an analysis of variance table is obtained for the regression model along with a table of the coefficients of the model, here the y-intercept (70.58) and the slope of the regression line (21.49).

In the results following the previous program, the Student's t-test associated with the slope can evaluate its significance in view of the data.

Adding a statement of the ODS GRAPHICS ON type to the REG procedure provides graphic outputs, including the scatter plot and the regression line (with confidence interval) superimposed. If it is desirable to distinctly show the observations according to the smoking status or not, it is possible to directly utilize graphic commands separated from the REG procedure. Here is an example with SGPLOT.

```
PROC SGPLOT  DATA=quetelet;
SCATTER x=qtt  y=pas / group=tab; REG x=qtt  y=pas ;
RUN;
```

To restrict the previous analyses to the smokers' group only (or non-smokers), we will proceed in exactly the same way but by prefixing the SAS instructions for the regression with a DATA stage, limiting the representations to the group of interest (tab variable), for example

```
DATA smoker; SET  quetelet;  IF  TAB=1;  RUN;
PROC REG DATA=smoker;  MODEL  pas=qtt;  RUN;
```

4) Based on the data about the birth weight ([HOS 89] and Chapter 1), we aim to study the relationship between the babies' weight (considered as a numeric variable, bwt) and two characteristics of the mother: her weight (lwt) and her ethnic origin (race).

a) Graphically represent the relationship between babies' weight and mothers' weight, according to the mothers' ethnicity.

b) Estimate the linear regression parameters considering the babies' weight as the response variable and the mothers' weight centered on their mean as the explanatory variable. Is the estimated slope significant at the usual 5 % threshold?

c) Estimate the parameters of the linear regression where this time the explanatory variable is the mothers' ethnicity and the response variable remains the same (babies' weight).

d) What is the predicted weight for a baby whose mother weighs 60 kg? Give a 95 % confidence interval for the mean predicted value.

The data can be imported, as previously, with Proc Import:

```
PROC IMPORT OUT= WORK.BIRTHWT
DATAFILE= "C:\folder\subfolder\Birthwt.txt"
DBMS=TAB REPLACE;
     GETNAMES=YES;
     DATAROW=2;
RUN;
```

Two solutions with SGPLOT here below:

```
PROC sort data= birthwt; BY race;
RUN;
PROC SGPLOT data= birthwt;
scatter x=bwt y=lwt;
reg x=bwt y=lwt;
by race;
RUN;
PROC SGPLOT data= birthwt;
scatter x=bwt y=lwt/group=race;
reg x=bwt y=lwt/group=race;
RUN;
```

These estimations can be obtained as following:

```
proc summary data=birthwt print mean;var lwt;output out=cl;run;
data cl1;set cl; if _STAT_ ne "MEAN" then delete;meanlwt=lwt;
keep meanlwt;run;
data new1;set cl1;
do i= 1 to  188;
output;
end;
drop i;
run;
data new;merge birthwt new1 ;
clwt=lwt-meanlwt;run;
PROC REG data=new;
model bwt=clwt;
run;
```

Ethnicity being a categorical variable, the REG procedure can no longer be utilized; PROC GLM applies.

```
PROC GLM data=new;
Class race;
model bwt=race;
run;
```

A possible solution is:

```
PROC GLM data=new;
model bwt=lwt ;
estimate "lwt" lwt 60;
run;
PROC GLM data=new;
model bwt=lwt/cli clm ;
run;
```

# 5

---

# Logistic Regression

---

In this chapter, we will introduce a new statistical procedure: PROC LOGISTIC.

PROC LOGISTIC is dedicated to logistic regression, a model in which the probability of an event associated with a response variable Y (disease occurrence) is expressed in a linear logistic manner according to a variable X (risk factor).

The regression coefficient is interpreted as a function of the odds ratio.

## 5.1. Logistic regression

### 5.1.1. Adjusted risk measure: Mantel-Haenszel statistics

```
PROC FREQ data=birthwt ;
TABLES race*smoke*low/CMH nopercent norow nocol;
run;
```

### 5.1.2. Simple logistic regression

```
PROC LOGISTIC  data=birthwt ;
MODEL low = smoke;
run;
```

| Frequency | Table 1 of smoke by low | | |
| --- | --- | --- | --- |
| | Controlling for race=1 | | |
| | | low | |
| smoke | 0 | 1 | Total |
| 0 | 40 | 4 | 44 |
| 1 | 33 | 19 | 52 |
| Total | 73 | 23 | 96 |

**Figure 5.1.** *Results from the execution of the FREQ procedure*

| Frequency | Table 2 of smoke by low | | |
| --- | --- | --- | --- |
| | Controlling for race=2 | | |
| | | low | |
| smoke | 0 | 1 | Total |
| 0 | 10 | 5 | 15 |
| 1 | 4 | 6 | 10 |
| Total | 14 | 11 | 25 |

**Figure 5.2.** *Continuation of the results from the execution of the FREQ procedure*

| Frequency | Table 3 of smoke by low | | |
| --- | --- | --- | --- |
| | Controlling for race=3 | | |
| | | low | |
| smoke | 0 | 1 | Total |
| 0 | 35 | 20 | 55 |
| 1 | 7 | 5 | 12 |
| Total | 42 | 25 | 67 |

**Figure 5.3.** *Continuation of the results from the execution of the FREQ procedure*

It is important to read the results.

By default, PROC LOGISTIC models the event corresponding to the first modality that appears. Therefore, here the event that is modeled is low=0. When the aim is to model the event low=1, the DESCENDING option will have to be introduced, thus:

```
PROC LOGISTIC data=birthwt DESCENDING;
MODEL low = smoke;
run;
```

| Cochran-Mantel-Haenszel Statistics (Based on Table Scores) | | | | |
|---|---|---|---|---|
| Statistic | Alternative Hypothesis | DF | Value | Prob |
| 1 | Nonzero Correlation | 1 | 9.0804 | 0.0026 |
| 2 | Row Mean Scores Differ | 1 | 9.0804 | 0.0026 |
| 3 | General Association | 1 | 9.0804 | 0.0026 |

| Common Odds Ratio and Relative Risks | | | | |
|---|---|---|---|---|
| Statistic | Method | Value | 95% Confidence Limits | |
| Odds Ratio | Mantel-Haenszel | 3.0316 | 1.4633 | 6.2810 |
| | Logit | 2.8852 | 1.3424 | 6.2010 |
| Relative Risk (Column 1) | Mantel-Haenszel | 1.3723 | 1.1107 | 1.6955 |
| | Logit | 1.3868 | 1.1338 | 1.6962 |
| Relative Risk (Column 2) | Mantel-Haenszel | 0.4721 | 0.2841 | 0.7846 |
| | Logit | 0.5541 | 0.3372 | 0.9106 |

| Breslow-Day Test for Homogeneity of the Odds Ratios | |
|---|---|
| Chi-Square | 3.1042 |
| DF | 2 |
| Pr > ChiSq | 0.2118 |

**Figure 5.4.** *Continuation of the results from the execution of the FREQ procedure*

| Analysis of Maximum Likelihood Estimates | | | | | |
|---|---|---|---|---|---|
| Parameter | DF | Estimate | Standard Error | Wald Chi-Square | Pr > ChiSq |
| Intercept | 1 | 1.0753 | 0.2150 | 25.0020 | <.0001 |
| smoke | 1 | -0.6924 | 0.3199 | 4.6855 | 0.0304 |

| Odds Ratio Estimates | | |
|---|---|---|
| Effect | Point Estimate | 95% Wald Confidence Limits |
| smoke | 0.500 | 0.267    0.937 |

| Association of Predicted Probabilities and Observed Responses | | | |
|---|---|---|---|
| Percent Concordant | 33.5 | Somers' D | 0.167 |
| Percent Discordant | 16.8 | Gamma | 0.333 |
| Percent Tied | 49.7 | Tau-a | 0.072 |
| Pairs | 7611 | c | 0.584 |

**Figure 5.5.** *Extract from the results from the execution of the LOGISTIC procedure*

The first modality read is 1 and therefore it is its probability that is modeled. The signs of the regression parameters change, the odds ratio is the inverse of the previous one and the rest remains unchanged.

## 5.1.3. *ROC curves*

```
DATA Data1;
    INPUT disease n age;
    CARDS;
  0  14 25
  0  20 35
  0  19 45
  7  18 55
  6  12 65
 17  17 75
  ;
```

```
PROC LOGISTIC data=Data1 plots(only)=roc(id=obs);
 MODEL disease/n=age/scale=none clparm=wald clodds=pl rsquare;
 units age=10;
 effectplot;
run;
```

| Analysis of Maximum Likelihood Estimates | | | | | |
|---|---|---|---|---|---|
| Parameter | DF | Estimate | Standard Error | Wald Chi-Square | Pr > ChiSq |
| Intercept | 1 | -1.0753 | 0.2150 | 25.0020 | <.0001 |
| smoke | 1 | 0.6924 | 0.3199 | 4.6855 | 0.0304 |

| Odds Ratio Estimates | | | |
|---|---|---|---|
| Effect | Point Estimate | 95% Wald Confidence Limits | |
| smoke | 1.998 | 1.068 | 3.741 |

**Figure 5.6.** *Extract from the results from the execution of the LOGISTIC procedure with the option DESCENDING*

The data that follow [DEL 88] focus on 49 patients with ovarian cancer also suffering from bowel obstruction. Three diagnostic tests are measured to determine whether the patient should have surgery.

```
data roc;
    input alb tp totscore popind @@;
    totscore = 10 - totscore;
    datalines;
3.0 5.8 10 0   3.2 6.3  5 1   3.9 6.8  3 1   2.8 4.8  6 0
3.2 5.8  3 1   0.9 4.0  5 0   2.5 5.7  8 0   1.6 5.6  5 1
3.8 5.7  5 1   3.7 6.7  6 1   3.2 5.4  4 1   3.8 6.6  6 1
4.1 6.6  5 1   3.6 5.7  5 1   4.3 7.0  4 1   3.6 6.7  4 0
2.3 4.4  6 1   4.2 7.6  4 0   4.0 6.6  6 0   3.5 5.8  6 1
3.8 6.8  7 1   3.0 4.7  8 0   4.5 7.4  5 1   3.7 7.4  5 1
```

```
3.1 6.6   6 1    4.1 8.2   6 1    4.3 7.0   5 1    4.3 6.5   4 1
3.2 5.1   5 1    2.6 4.7   6 1    3.3 6.8   6 0    1.7 4.0   7 0
3.7 6.1   5 1    3.3 6.3   7 1    4.2 7.7   6 1    3.5 6.2   5 1
2.9 5.7   9 0    2.1 4.8   7 1    2.8 6.2   8 0    4.0 7.0   7 1
3.3 5.7   6 1    3.7 6.9   5 1    3.6 6.6   5 1
;
proc logistic data=roc plots=roc(id=prob);
   model popind(event='0') = alb tp totscore / nofit;
   roc 'Albumin' alb;
   roc 'K-G Score' totscore;
   roc 'Total Protein' tp;
   roccontrast reference('K-G Score') / estimate e;
run;
```

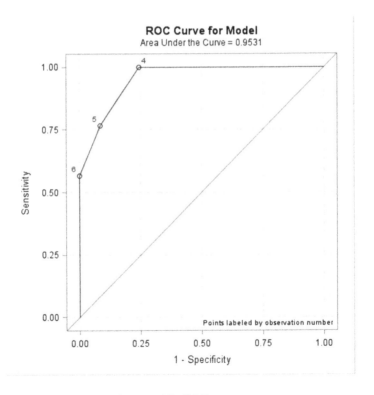

**Figure 5.7.** *ROC curve*

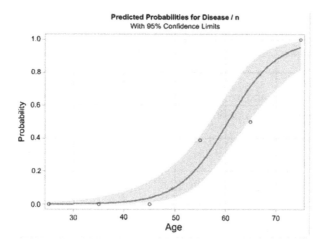

**Figure 5.8.** *Continuation of the results from the
previous logistic procedure*

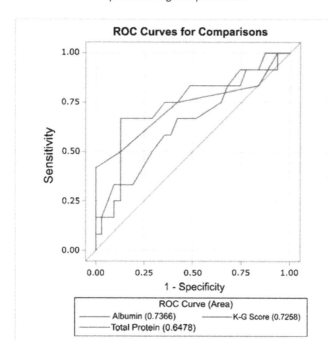

**Figure 5.9.** *Several ROC curves on the same graph. For a color
version of this figure, see www.iste.co.uk/lalanne/biostatistics3.zip*

## 5.2. Key points to remember

In this chapter, we have essentially presented the PROC LOGISTIC procedure:

– the odds ratio and relative risks of the Mantel-Haenszel method are obtained with the CMH option of the "tables" statement of PROC FREQ;

– the parameters of a single or multiple logistic model can be estimated with PROC LOGISTIC;

– the option plots=roc() and the ROC statement enable a ROC curve to be generated;

– the statement roccontrast allows several ROC curves to be compared.

| ROC Association Statistics | | | | | | | |
|---|---|---|---|---|---|---|---|
| | | | Mann-Whitney | | | | |
| ROC Model | Area | Standard Error | 95% Wald Confidence Limits | | Somers' D (Gini) | Gamma | Tau-a |
| Albumin | 0.7366 | 0.0927 | 0.5549 | 0.9182 | 0.4731 | 0.4809 | 0.1949 |
| K-G Score | 0.7258 | 0.1028 | 0.5243 | 0.9273 | 0.4516 | 0.5217 | 0.1860 |
| Total Protein | 0.6478 | 0.1000 | 0.4518 | 0.8439 | 0.2957 | 0.3107 | 0.1218 |

| ROC Contrast Coefficients | | |
|---|---|---|
| ROC Model | Row1 | Row2 |
| Albumin | 1 | 0 |
| K-G Score | -1 | -1 |
| Total Protein | 0 | 1 |

Figure 5.10. *Comparison of ROC curves*

## 5.3. Further information

Again, the help and the examples in the online usage manual of SAS can further assist the reader in finding applications for the LOGISTIC procedure which are not described in this book.

Logistic regression models are also carried out by using PROC GLIMMIX, which deals with random factors and random coefficients, not covered in this book.

| ROC Contrast Test Results | | | |
|---|---|---|---|
| Contrast | DF | Chi-Square | Pr > ChiSq |
| Reference = K-G Score | 2 | 2.5340 | 0.2817 |

| ROC Contrast Estimation and Testing Results by Row | | | | | | |
|---|---|---|---|---|---|---|
| Contrast | Estimate | Standard Error | 95% Wald Confidence Limits | | Chi-Square | Pr > ChiSq |
| Albumin - K-G Score | 0.0108 | 0.0953 | -0.1761 | 0.1976 | 0.0127 | 0.9102 |
| Total Protein - K-G Score | -0.0780 | 0.1046 | -0.2830 | 0.1271 | 0.5554 | 0.4561 |

**Figure 5.11.** *Comparison of ROC curves (continuation)*

## 5.4. Applications

1) We study the effect of prophylactic therapy of a macrolide with low doses (treatment A) on the infectious episodes in patients suffering from cystic fibrosis in a placebo-controlled multicenter randomized trial (B). The results are summarized in Table 5.1.

a) Based on a $\chi^2$ test, how can the following question be addressed: can the treatment prevent the occurrence of infectious episodes (at threshold $\alpha = 0.05$)? Verify that all the expected counts are effectively larger than 5.

b) Can the same conclusion be achieved considering the confidence interval of the odds ratio associated with the treatment effect?

c) We want to verify whether there is a disparity from the point of view of the percentages of infectious episodes according to the center. The data per center are indicated in the table hereafter. Draw a conclusion based on the $\chi^2$ test.

d) Based on the previous table, we want to verify whether the treatment effect is independent or not of the center. A possible approach is to carry out a comparison test between the two treatments adjusted on the center (Mantel-Haenszel test). Indicate the result of the test as well as the value of the adjusted odds ratio.

| | Infection | | |
|---|---|---|---|
| | No | Yes | Total |
| Treatment (A) | 157 | 52 | 209 |
| Placebo (B) | 119 | 103 | 222 |
| Total | 276 | 155 | 431 |

**Table 5.1.** *Prophylactic and cystic fibrosis treatment*

| | Infection | | | | Infection | | | | Infection | | |
|---|---|---|---|---|---|---|---|---|---|---|---|
| | No | Yes | Total | | No | Yes | Total | | No | Yes | Total |
| Treatment (A) | 51 | 8 | 59 | Treatment (A) | 91 | 35 | 126 | Treatment (A) | 15 | 9 | 24 |
| Placebo (B) | 47 | 19 | 66 | Placebo (B) | 61 | 71 | 132 | Placebo (B) | 11 | 13 | 24 |
| Total | 98 | 27 | 125 | Total | 152 | 106 | 258 | Total | 26 | 22 | 48 |
| | Center 1 | | | | Center 2 | | | | Center 3 | | |

First, it is necessary to build the count table given in the statement. This can be achieved by means of a DATA step as described below.

```
DATA mucovisi;
INPUT infection treatment number;
cards;
0  1   157
1  1   52
0  0   119
1  0   103
; RUN;
PROC FORMAT;  VALUE yesno   0="No"   1="Yes";
VALUE treat  0="Placebo  (B)"  1="Treatment  (A)";
RUN;
```

It should be noted that more informative labels have been added to the numerical codes used to represent the modalities of the two binary variables, infection and treatment.

Here follows a way to reproduce the table from the wording with the associated $\chi^2$ test (chisq option). It should be noted that the WEIGHT statement is used to indicate to SAS that the data are grouped and that the table should be weighted with the counts reported in the 3rd column of the data (named number during the DATA step).

```
PROC FREQ  DATA=mucovisi   ORDER=data;
TABLES   treatment*infection/expected   chisq;
```

```
WEIGHT   number;
FORMAT   infection   yesno.   treatment   treat.;
RUN;
```

It should be noted that the observed and theoretical counts are presented in the same table by SAS.

By adding the relrisk option to the TABLES statement, we obtain risk measures typically used in epidemiology.

```
PROC FREQ   DATA=mucovisi   ORDER=data;
TABLES   treatment*infection/expected chisq relrisk;
WEIGHT   number;
FORMAT   infection   yesno.   treatment   treat.;
RUN;
```

When taking into account the center, it is necessary to rebuild the count table, considering only the column margins of the count tables given in the wording. The following shows how to proceed with SAS.

```
DATA mucocenters;
INPUT   center   infection   treatment number;
cards;
1  0  1  51
1  1  1  8
1  0  0  47
1  1  0  19
2  0  1  91
2  1  1  35
2  0  0  61
2  1  0  71
3  0  1  15
3  1  1  9
3  0  0  11
3  1  0  13
; RUN;
```

To perform the Mantel-Haenszel test, we use the totality of the data ($3\ 2 \times 2$ tables) and PROC FREQ.

```
PROC FREQ   DATA=mucocenters   ORDER=data;
TABLES   center*treatment*infection/cmh; WEIGHT   number;
FORMAT   infection   yesno.   treatment   treat.; RUN;
```

2) Table 5.2 summarizes the proportion of myocardial infarction among men aged from 40 to 59 years old and for whom the blood pressure level and the cholesterol rate have been registered, considered in the form of ordered classes [EVE 01].

| TA | Cholesterol (mg/100 ml) | | | | | | |
|---|---|---|---|---|---|---|---|
| | < 200 | 200 − 209 | 210 − 219 | 220 − 244 | 245 − 259 | 260 − 284 | > 284 |
| < 117 | 2/53 | 0/21 | 0/15 | 0/20 | 0/14 | 1/22 | 0/11 |
| 117 − 126 | 0/66 | 2/27 | 1/25 | 8/69 | 0/24 | 5/22 | 1/19 |
| 127 − 136 | 2/59 | 0/34 | 2/21 | 2/83 | 0/33 | 2/26 | 4/28 |
| 137 − 146 | 1/65 | 0/19 | 0/26 | 6/81 | 3/23 | 2/34 | 4/23 |
| 147 − 156 | 2/37 | 0/16 | 0/6 | 3/29 | 2/19 | 4/16 | 1/16 |
| 157 − 166 | 1/13 | 0/10 | 0/11 | 1/15 | 0/11 | 2/13 | 4/12 |
| 167 − 186 | 3/21 | 0/5 | 0/11 | 2/27 | 2/5 | 6/16 | 3/14 |
| > 186 | 1/5 | 0/1 | 3/6 | 1/10 | 1/7 | 1/7 | 1/7 |

**Table 5.2.** *Infarction and blood pressure*

The data are available in the file hdis.dat in the form of a table comprising four columns respectively indicating blood pressure (eight categories, denoted by 1 to 8), cholesterol rate (seven categories, denoted by 1 to 7), the number of heart attacks and the total number of individuals. We are interested in the association between blood pressure and the likelihood of having a heart attack.

a) Calculate the proportions of heart attack for every level of blood pressure and represent them in a table and in graphic form.

b) Express the proportions calculated in (a) in the form of logit.

c) Based on a logistic regression model, determine whether there is a significant association at the threshold $\alpha = 0.05$ between blood pressure, assumed as a quantitative variable considering class centers, and the likelihood of having a heart attack.

d) Express in logit units, the probabilities of heart attack predicted by the model for each of the blood pressure levels.

e) On the same graph, display empirical proportions and the logistic regression curve based on blood pressure values (class centers).

The file can be imported with:

```
PROC IMPORT  OUT= WORK.hdis
DATAFILE=  "C:\folder\subfolder\hdis.dat" DBMS=DLM  REPLACE;
DELIMITER='   'x; GETNAMES=YES;
DATAROW=2;
RUN;
```

Proc format and the following data steps enable a presentation of results in accordance with the table in the statement.

```
Proc format;
value Pressure 1='< 117' 2='117 - 126' 3='127 - 136'
    4='137 - 146' 5='147 - 156' 6='157 - 166' 7='167 - 186'
    8='> 186';
value Cholest 1='< 200' 2='200 - 209' 3='210 - 219'
    4='220 - 244' 5='244 - 259' 6='260 - 284' 7='> 284';
run;
data hdis1;set hdis;
label bpress ='TA';
label chol ='Cholesterol (mg/100ml)';
label hdis='Myocardial Infarction';
run;
```

The calculation of the proportions of heart attack for every blood pressure level is done utilizing sequences of Proc freq and data steps:

```
proc freq data=hdis1 noprint;
tables bpress*chol/nopercent norow nocol out=File1 sparse;
weight hdis;
format bpress Pressure. chol Cholest.;
run;
data file1;set file1; count1=count;drop count percent;
label count1='Number of heart attacks';run;
run;
```

The File1 file contains the contingency table of cases (number of heart attacks). Then:

```
proc freq data=hdis1 noprint;
tables bpress*chol/nopercent norow nocol out=File2 sparse;
weight total;
format bpress Pressure. chol Cholest.;
run;
data file2;set file2; count2=count;drop count percent;
label count2='Total number';run;
```

The File2 file contains the contingency table of totals by crossing (total number). Then, using the MERGE statement will make it possible to constitute the file in which the proportions of heart attacks will be calculated.

```
data file;merge file1 file2;
Proportion=count1/count2;
logitP=exp(Proportion)/(1+exp(Proportion));
LogitP_=logistic(Proportion);
bpress_q=110; if bpress=2 then bpress_q=121.5;
else if bpress=3 then bpress_q=131.5;
else if  bpress=4 then bpress_q=141.5;
else if bpress=5 then bpress_q=151.5;
else if bpress=6 then bpress_q=161.5;
else if bpress=7 then bpress_q=171.5;
else if bpress=8 then bpress_q=190;
label Proportion ='Proportion of heart attacks';
label LogitP='Proportion of heart attacks in logit';
label LogitP_='Proportion of heart attacks in logit';
label bpress_q='TA quantitative';
run;
```

The contingency table of the cases is thus displayed:

```
proc freq data=file; tables bpress*chol/nopercent norow nocol;
weight count1;
format bpress Pressure. chol Cholest.;
run;
```

That of totals:

```
proc freq data=file; tables bpress*chol/nopercent norow nocol;
weight count2;
format bpress Pressure. chol Cholest.;
run;
```

And finally, that of proportions of heart attack:

```
proc freq data=file; tables bpress*chol/nopercent norow nocol;
weight proportion;
format bpress Pressure. chol Cholest.;
run;
```

And in graphic form:

```
symbol i=stdm ;
proc gplot data=file;
   plot proportion*bpress;
      format bpress Pressure.;
run;
```

The calculation of proportions in the form of logit was carried out in the previous data step, in two ways: by using the logit expression as the exponential function and by using the logistic function, which is programmed in SAS.

The quantitative variable, class center, has also been created in the previous data step ($bpress_q$). The following program makes use of proc logistic and the model includes the quantitative variable $bress_q$, the variable chol, regarded as ordinal and their interaction.

```
proc logistic data=file;
class chol/param=ordinal;
model count1  / count2=bpress_q chol bpress_q*chol ;
output out=fic1 p=probabilit;
format chol Cholest.;
run;
```

The interaction not being significant, the following model, without interaction, will be executed:

```
proc logistic data=file;
class chol/param=ordinal;
model count1  / count2=bpress_q chol  ;
output out=fic2 p=probabilit xbeta=logitpred;
format chol Cholest.;
run;
```

The expression in logit units of the probabilities of heart attacks predicted by the model for each of the blood pressure levels is obtained in the previous program as column (logitpred) of the fic2 file, by means of the option xbeta=logitpred.

Displayed on the same graph, empirical proportions and the logistic regression curve according to blood pressure values (class centers) are obtained by means of the following program:

```
symbol i=stdmj;
proc gplot data=fic2;
  plot probabilit*bpress proportion*bpress/overlay;
  format bpress Pressure.;
run;
```

3) A case-control investigation has focused on the relationship between alcohol and tobacco consumption and esophageal cancer in humans (study "Ille and Villaine"). The group of cases was constituted of 200 patients suffering from esophagus cancer and diagnosed between January 1972 and April 1974. In total, 775 male controls have been selected from the electoral lists. Table 5.3 shows the distribution of all the subjects according to their daily consumption of alcohol, bearing in mind that alcohol consumption greater than 80 g is considered to be a risk factor [BRE 80]. The data are available in the file cc_oesophage.csv.

a) What is the value of the odds ratio and its 95% confidence interval (Woolf method)? Is this a good estimate of the relative risk?

b) Is the proportion of consumers at risk the same in cases and in controls (consider $\alpha = 0.05$)?

c) Build the logistic regression model for testing the association between alcohol consumption and subjects' status. Is the regression coefficient significant?

d) Recover the value of the observed odds ratio, calculated in (a), and its confidence interval based on the results of the regression analysis.

| | Alcohol intake (g/day) | | Total |
|---|---|---|---|
| | ≥ 80 | < 80 | |
| Cases | 96 | 104 | 200 |
| Controls | 109 | 666 | 775 |
| Total | 205 | 770 | 975 |

**Table 5.3.** *Infarction and blood pressure*

The data have been saved in a compact format (three columns indicating the presence or the lack of cancer, the level of alcohol consumption and the associated counts). The following shows a way to store the data table in SAS.

```
DATA alcohol;
INPUT  Consum  Cancer  number;
cards;
1  1  96
0  1  104
1  0  109
0  0  666
; RUN;

PROC FORMAT;  VALUE alcohol  1=">=  80" 0="<80";
VALUE case 1="Case"  0="Controls" ;
RUN;
```

The proportion of individuals at risk, that is having a daily consumption of alcohol = 80 g, is obtained from a simple count table crossing variables cancer and alcohol (it is necessary to indicate how the cells have to be filled by adding a weight option). SAS will automatically provide the row profiles, that is the relative frequencies per row.

```
PROC FREQ  DATA=alcohol;  TABLES  cancer*consum/ RELRISK;
WEIGHT  number;
FORMAT  consum alcohol.  cancer  case.;
RUN;
```

The question is addressed in the previous program.

The logistic regression model is formulated as follows under SAS; once again, a WEIGHT option indicates the weighting with the counts of the grouped data.

```
PROC LOGISTIC  DATA=alcohol;
MODEL  Cancer  =  consum;
WEIGHT  number;
RUN;
```

# 6

# Survival Curves, Cox Regression

In this final chapter, we will essentially introduce two new statistical procedures: PROC LIFETEST and PROC PHREG. These two procedures enable the analysis of survival data.

These survival data are periods of time which have the peculiarity that some of them are not fully observed.

These periods of time are usually the length of time between two events: a birth and a death. Some of these times are censored: the real time is not observed, because the date of death is not known. The study was completed before the death of some people, or, some people abandoned the study at some point, so, their real status (living or deceased) is unknown.

In summary, in this type of study, the response variable is a pair consisting of two variables: a positive quantitative variable (the time period) and a variable indicating whether this time corresponds to a death (or more generally the occurrence of an event of interest) or to censoring (and this is referred to as observation time).

PROC LIFETEST makes it possible to estimate the probability Prob $(S > t)$, called the survival function, and to perform simple tests for comparing survival functions within groups.

PROC PHREG is a means to perform more complex analyses on these lifetimes. It enables the estimation of Cox regression models, the testing of assumptions about its parameters and testing the assumptions underlying this model (proportional hazards assumption).

## 6.1. Survival curves

### 6.1.1. Kaplan-Meier method

In the following example, the lifetimes of 137 patients who have received a bone marrow transplant have been registered. At the time of the transplant, each patient is part of a risk group of a possible three: ALL (Acute Lymphoblastic Leukemia), low risk AML (Acute Myelocytic Leukemia) and high risk AML. The variable T represents the time period until death, until a relapse or until the end of the study. The Status variable is a censoring indicator (1 = event, 0 = censoring).

```
proc format;
    value risk 1='ALL' 2='AML-Low Risk' 3='AML-High Risk';
data BMT;
    input Group T Status @@;
    format Group risk.;
    label T='Disease Free Time';
    datalines;
1 2081 0 1 1602 0 1 1496 0 1 1462 0 1 1433 0
1 1377 0 1 1330 0 1  996 0 1  226 0 1 1199 0
1 1111 0 1  530 0 1 1182 0 1 1167 0 1  418 1
1  383 1 1  276 1 1  104 1 1  609 1 1  172 1
1  487 1 1  662 1 1  194 1 1  230 1 1  526 1
1  122 1 1  129 1 1   74 1 1  122 1 1   86 1
1  466 1 1  192 1 1  109 1 1   55 1 1    1 1
1  107 1 1  110 1 1  332 1 2 2569 0 2 2506 0
2 2409 0 2 2218 0 2 1857 0 2 1829 0 2 1562 0
2 1470 0 2 1363 0 2 1030 0 2  860 0 2 1258 0
2 2246 0 2 1870 0 2 1799 0 2 1709 0 2 1674 0
2 1568 0 2 1527 0 2 1324 0 2  957 0 2  932 0
2  847 0 2  848 0 2 1850 0 2 1843 0 2 1535 0
2 1447 0 2 1384 0 2  414 1 2 2204 1 2 1063 1
2  481 1 2  105 1 2  641 1 2  390 1 2  288 1
2  421 1 2   79 1 2  748 1 2  486 1 2   48 1
```

```
2   272 1 2 1074 1 2   381 1 2    10 1 2    53 1
2    80 1 2    35 1 2   248 1 2   704 1 2   211 1
2   219 1 2   606 1 3  2640 0 3  2430 0 3  2252 0
3  2140 0 3  2133 0 3  1238 0 3  1631 0 3  2024 0
3  1345 0 3  1136 0 3   845 0 3   422 1 3   162 1
3    84 1 3   100 1 3     2 1 3    47 1 3   242 1
3   456 1 3   268 1 3   318 1 3    32 1 3   467 1
3    47 1 3   390 1 3   183 1 3   105 1 3   115 1
3   164 1 3    93 1 3   120 1 3    80 1 3   677 1
3    64 1 3   168 1 3    74 1 3    16 1 3   157 1
3   625 1 3    48 1 3   273 1 3    63 1 3    76 1
3   113 1 3   363 1
;
run;
PROC LIFETEST DATA=BMT ;TIME T * Status(0);run;
```

### The LIFETEST Procedure

| | Product-Limit Survival Estimates | | | | |
|---|---|---|---|---|---|
| T | Survival | Failure | Survival Standard Error | Number Failed | Number Left |
| 0.00 | 1.0000 | 0 | 0 | 0 | 137 |
| 1.00 | 0.9927 | 0.00730 | 0.00727 | 1 | 136 |
| 2.00 | 0.9854 | 0.0146 | 0.0102 | 2 | 135 |
| 10.00 | 0.9781 | 0.0219 | 0.0125 | 3 | 134 |
| 16.00 | 0.9708 | 0.0292 | 0.0144 | 4 | 133 |
| 32.00 | 0.9635 | 0.0365 | 0.0160 | 5 | 132 |
| 35.00 | 0.9562 | 0.0438 | 0.0175 | 6 | 131 |
| 47.00 | . | . | . | 7 | 130 |
| 47.00 | 0.9416 | 0.0584 | 0.0200 | 8 | 129 |
| 48.00 | . | . | . | 9 | 128 |
| 48.00 | 0.9270 | 0.0730 | 0.0222 | 10 | 127 |
| 53.00 | 0.9197 | 0.0803 | 0.0232 | 11 | 126 |
| 55.00 | 0.9124 | 0.0876 | 0.0242 | 12 | 125 |
| 63.00 | 0.9051 | 0.0949 | 0.0250 | 13 | 124 |

**Figure 6.1.** *Results from the execution of the lifetest procedure*

For obvious reasons, the tables in Figures 6.1 and 6.2 are shown only partially.

| 1850.00 | * | . | . | . | 82 | 15 |
|---|---|---|---|---|---|---|
| 1857.00 | * | . | . | . | 82 | 14 |
| 1870.00 | * | . | . | . | 82 | 13 |
| 2024.00 | * | . | . | . | 82 | 12 |
| 2081.00 | * | . | . | . | 82 | 11 |
| 2133.00 | * | . | . | . | 82 | 10 |
| 2140.00 | * | . | . | . | 82 | 9 |
| 2204.00 | | 0.3509 | 0.6491 | 0.0559 | 83 | 8 |
| 2218.00 | * | . | . | . | 83 | 7 |
| 2246.00 | * | . | . | . | 83 | 6 |
| 2252.00 | * | . | . | . | 83 | 5 |
| 2409.00 | * | . | . | . | 83 | 4 |
| 2430.00 | * | . | . | . | 83 | 3 |
| 2506.00 | * | . | . | . | 83 | 2 |
| 2569.00 | * | . | . | . | 83 | 1 |
| 2640.00 | * | 0.3509 | 0.6491 | . | 83 | 0 |

Note: The marked survival times are censored observations.

**Figure 6.2.** *Continuation of the results from the execution of the lifetest procedure*

This is the simplest use of PROC LIFETEST. Kaplan-Meier estimations of survival probability, in addition to likelihoods of death, are given every time an event occurs. The estimate of the median survival time (here 481), those corresponding to quartiles 0.25 and 0.75, the mean lifetime are given by default. The Kaplan-Meier curve representing the previous survival probabilities according to the lifetime is also given.

Summary Statistics for Time Variable T

| Quartile Estimates | | | | |
|---|---|---|---|---|
| | Point | 95% Confidence Interval | | |
| Percent | Estimate | Transform | [Lower | Upper) |
| 75 | . | LOGLOG | 2204.00 | . |
| 50 | 481.00 | LOGLOG | 363.00 | 748.00 |
| 25 | 122.00 | LOGLOG | 104.00 | 194.00 |

| Mean | Standard Error |
|---|---|
| 1033.09 | 83.37 |

**Figure 6.3.** *Continuation of the results from the lifetest procedure: survival median and mean*

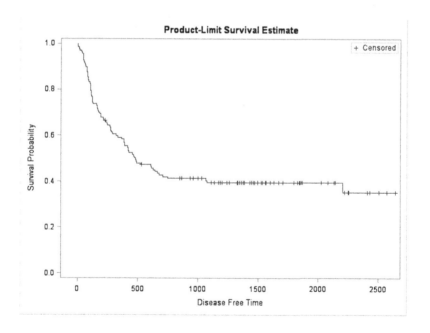

**Figure 6.4.** *Continuation of the results from the lifetest procedure: Kaplan-Meier curve*

| Summary of the Number of Censored and Uncensored Values | | | |
|---|---|---|---|
| Total | Failed | Censored | Percent Censored |
| 137 | 83 | 54 | 39.42 |

**Figure 6.5.** *Continuation of the results from the lifetest procedure: number of censored observations*

## 6.1.2. *Log-rank test: comparison of the survival of several populations*

The PROC LIFETEST program hereafter compares the survival curves of three populations defined by the variable Group. Again, for obvious reasons, the tables in Figures 6.5, 6.7 and 6.9 are shown only partially.

```
PROC LIFETEST data=BMT PLOTS = SURVIVAL(ATRISK=0 to 2500 by 500);
   TIME T * Status(0);
   STRATA Group / TEST= LOGRANK ADJUST=SIDAK;
   run;
```

The LIFETEST Procedure

Stratum 1: Group = ALL

| T | Survival | Failure | Survival Standard Error | Number Failed | Number Left |
|---|---|---|---|---|---|
| | | Product-Limit Survival Estimates | | | |
| 0.00 | 1.0000 | 0 | 0 | 0 | 38 |
| 1.00 | 0.9737 | 0.0263 | 0.0260 | 1 | 37 |
| 55.00 | 0.9474 | 0.0526 | 0.0362 | 2 | 36 |
| 74.00 | 0.9211 | 0.0789 | 0.0437 | 3 | 35 |
| 86.00 | 0.8947 | 0.1053 | 0.0498 | 4 | 34 |

**Figure 6.6.** *Results from the execution of the lifetest procedure: log-rank test*

**Summary Statistics for Time Variable T**

| Quartile Estimates | | | | |
|---|---|---|---|---|
| | Point | 95% Confidence Interval | | |
| Percent | Estimate | Transform | [Lower | Upper) |
| 75 | . | LOGLOG | 609.00 | . |
| 50 | 418.00 | LOGLOG | 192.00 | . |
| 25 | 122.00 | LOGLOG | 86.00 | 230.00 |

| Mean | Standard Error |
|---|---|
| 398.24 | 41.11 |

**Figure 6.7.** *Continuation of the results from the lifetest procedure: log-rank test. Note: The mean survival time and its standard error were underestimated because the largest observation was censored and the estimation was restricted to the largest event time*

The following program differs from the previous one by the graph of survival curves, with confidence regions (Figure 6.15).

```
PROC LIFETEST data=BMT PLOTS = SURVIVAL(ATRISK=0 to 2500 by 500 CL);
    TIME T * Status(0);
    STRATA Group / TEST= LOGRANK ADJUST=SIDAK;
    run;
```

## 6.2. Cox regression

### 6.2.1. *Cox regression, one independent variable*

Hereafter, a simple PROC PHREG program, with one categorical variable, the variable GROUP as an independent variable.

```
PROC PHREG data=BMT;
CLASS GROUP;
MODEL T*STATUS(0)= GROUP;
RUN;
```

Stratum 2: Group = AML-High Risk

| | Product-Limit Survival Estimates | | | | |
|---|---|---|---|---|---|
| T | Survival | Failure | Survival Standard Error | Number Failed | Number Left |
| 0.00 | 1.0000 | 0 | 0 | 0 | 45 |
| 2.00 | 0.9778 | 0.0222 | 0.0220 | 1 | 44 |
| 16.00 | 0.9556 | 0.0444 | 0.0307 | 2 | 43 |
| 32.00 | 0.9333 | 0.0667 | 0.0372 | 3 | 42 |
| 47.00 | . | . | . | 4 | 41 |

**Figure 6.8.** *Continuation of the results from the lifetest procedure: log-rank test*

Summary Statistics for Time Variable T

| | Quartile Estimates | | | |
|---|---|---|---|---|
| | Point | | 95% Confidence Interval | |
| Percent | Estimate | Transform | [Lower | Upper) |
| 75 | 677.00 | LOGLOG | 363.00 | . |
| 50 | 183.00 | LOGLOG | 113.00 | 390.00 |
| 25 | 84.00 | LOGLOG | 48.00 | 115.00 |

| Mean | Standard Error |
|---|---|
| 312.47 | 38.69 |

**Figure 6.9.** *Continuation of the results from the lifetest procedure: log-rank test. Note: The mean survival time and its standard error were underestimated because the largest observation was censored and the estimation was restricted to the largest event time*

## Stratum 3: Group = AML-Low Risk

| | Product-Limit Survival Estimates | | | | |
|---|---|---|---|---|---|
| T | Survival | Failure | Survival Standard Error | Number Failed | Number Left |
| 0.00 | 1.0000 | 0 | 0 | 0 | 54 |
| 10.00 | 0.9815 | 0.0185 | 0.0183 | 1 | 53 |
| 35.00 | 0.9630 | 0.0370 | 0.0257 | 2 | 52 |
| 48.00 | 0.9444 | 0.0556 | 0.0312 | 3 | 51 |
| 53.00 | 0.9259 | 0.0741 | 0.0356 | 4 | 50 |

**Figure 6.10.** *Continuation of the results from the lifetest procedure: log-rank test*

## Summary Statistics for Time Variable T

| | Quartile Estimates | | | |
|---|---|---|---|---|
| | Point | 95% Confidence Interval | | |
| Percent | Estimate | Transform | [Lower | Upper) |
| 75 | . | LOGLOG | . | . |
| 50 | 2204.00 | LOGLOG | 641.00 | . |
| 25 | 390.00 | LOGLOG | 105.00 | 641.00 |

| Mean | Standard Error |
|---|---|
| 1382.46 | 129.65 |

**Figure 6.11.** *Continuation of the results from the lifetest procedure: log-rank test. Note: The mean survival time and its standard error were underestimated because the largest observation was censored and the estimation was restricted to the largest event time*

| Summary of the Number of Censored and Uncensored Values | | | | | |
|---|---|---|---|---|---|
| Stratum | Group | Total | Failed | Censored | Percent Censored |
| 1 | ALL | 38 | 24 | 14 | 36.84 |
| 2 | AML-High Risk | 45 | 34 | 11 | 24.44 |
| 3 | AML-Low Risk | 54 | 25 | 29 | 53.70 |
| Total | | 137 | 83 | 54 | 39.42 |

**Figure 6.12.** *Continuation of the results from the lifetest procedure: log-rank test*

**Testing Homogeneity of Survival Curves for T over Strata**

| Rank Statistics | |
|---|---|
| Group | Log-Rank |
| ALL | 2.148 |
| AML-High Risk | 12.818 |
| AML-Low Risk | -14.966 |

| Covariance Matrix for the Log-Rank Statistics | | | |
|---|---|---|---|
| Group | ALL | AML-High Risk | AML-Low Risk |
| ALL | 15.9552 | -5.6101 | -10.3451 |
| AML-High Risk | -5.6101 | 15.6048 | -9.9947 |
| AML-Low Risk | -10.3451 | -9.9947 | 20.3398 |

| Test of Equality over Strata | | | |
|---|---|---|---|
| Test | Chi-Square | DF | Pr > Chi-Square |
| Log-Rank | 13.8037 | 2 | 0.0010 |

**Figure 6.13.** *Continuation of the results from the lifetest procedure: log-rank test*

| Adjustment for Multiple Comparisons for the Logrank Test | | | | |
|---|---|---|---|---|
| Strata Comparison | | | p-Values | |
| Group | Group | Chi-Square | Raw | Sidak |
| ALL | AML-High Risk | 2.6610 | 0.1028 | 0.2779 |
| ALL | AML-Low Risk | 5.1400 | 0.0234 | 0.0685 |
| AML-High Risk | AML-Low Risk | 13.8011 | 0.0002 | 0.0006 |

**Figure 6.14.** *Continuation of the results from the lifetest procedure: log-rank test*

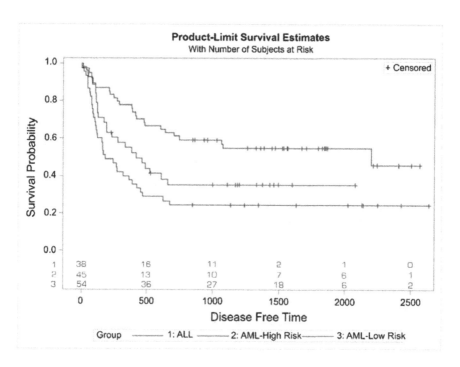

**Figure 6.15.** *Survival curves with number of subjects at risk. For a color version of this figure, see www.iste.co.uk/lalanne/biostatistics3.zip*

The following program differs from the previous one by the graph of cumulative hazards survival curves, with confidence regions (Figure 6.15).

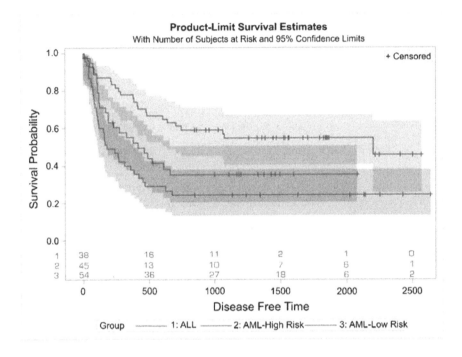

**Figure 6.16.** *Confidence regions for survival curves. For a color version of this figure, see www.iste.co.uk/lalanne/biostatistics3.zip*

```
PROC PHREG data=BMT PLOTS (CL)=(survival cumhaz);
CLASS GROUP;
MODEL T*STATUS(0)= GROUP;
RUN;
```

The following PROC PHREG program will test the proportional hazards assumption for the independent variable group considered as ordinal (no CLASS statement).

```
PROC PHREG data=BMT;
MODEL T*STATUS(0)= GROUP;
ASSESS VAR=(GROUP) PH/ RESAMPLE;
RUN;
```

The PHREG Procedure

| Model Information | | |
|---|---|---|
| Data Set | WORK.BMT | |
| Dependent Variable | T | Disease Free Time |
| Censoring Variable | Status | |
| Censoring Value(s) | 0 | |
| Ties Handling | BRESLOW | |

| | |
|---|---|
| Number of Observations Read | 137 |
| Number of Observations Used | 137 |

| Class Level Information | | | |
|---|---|---|---|
| Class | Value | Design Variables | |
| Group | ALL | 1 | 0 |
| | AML-High Risk | 0 | 1 |
| | AML-Low Risk | 0 | 0 |

| Summary of the Number of Event and Censored Values | | | |
|---|---|---|---|
| Total | Event | Censored | Percent Censored |
| 137 | 83 | 54 | 39.42 |

**Figure 6.17.** *Results from the PHREG procedure*

The ASSESS statement with the PH and RESAMPLE options allows the goodness-of-fit test of the proportional hazards assumption to be performed. The linear functional relationship between the response and the group (variable considered as ordinal) is also tested.

### 6.2.2. Cox regression, two independent variables

The data have been collected from 59 patients suffering from breast cancer and are available in the file polymorphism.dta (Stata file) [DUP 09]. Mean data are indicated in Table 3.2. Data in Stata format can be imported into SAS by using PROC IMPORT and indicating the type of data source, here DBMS=STATA.

| Model Fit Statistics | | |
|---|---|---|
| Criterion | Without Covariates | With Covariates |
| -2 LOG L | 746.719 | 733.288 |
| AIC | 746.719 | 737.288 |
| SBC | 746.719 | 742.126 |

| Testing Global Null Hypothesis: BETA=0 | | | |
|---|---|---|---|
| Test | Chi-Square | DF | Pr > ChiSq |
| Likelihood Ratio | 13.4307 | 2 | 0.0012 |
| Score | 13.7821 | 2 | 0.0010 |
| Wald | 13.0092 | 2 | 0.0015 |

| Type 3 Tests | | | |
|---|---|---|---|
| Effect | DF | Wald Chi-Square | Pr > ChiSq |
| Group | 2 | 13.0092 | 0.0015 |

| Analysis of Maximum Likelihood Estimates | | | | | | | | |
|---|---|---|---|---|---|---|---|---|
| Parameter | | DF | Parameter Estimate | Standard Error | Chi-Square | Pr > ChiSq | Hazard Ratio | Label |
| Group | ALL | 1 | 0.57418 | 0.28730 | 3.9942 | 0.0457 | 1.776 | Group ALL |
| Group | AML-High Risk | 1 | 0.95680 | 0.26535 | 13.0019 | 0.0003 | 2.603 | Group AML-High Risk |

**Figure 6.18.** *Continuation of the results from the PHREG procedure*

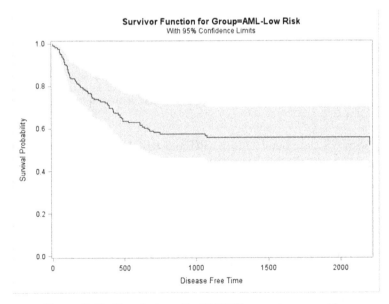

**Figure 6.19.** *Results from the PHREG procedure, graphics*

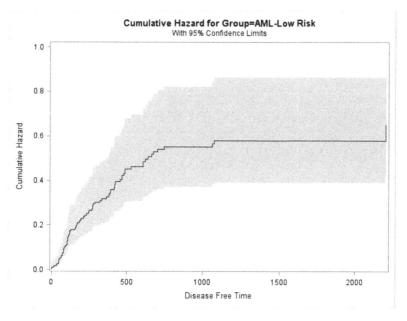

**Figure 6.20.** *Continuation of the results from the
PHREG procedure, graphics*

| Analysis of Maximum Likelihood Estimates | | | | | | |
|---|---|---|---|---|---|---|
| Parameter | DF | Parameter Estimate | Standard Error | Chi-Square | Pr > ChiSq | Hazard Ratio |
| Group | 1 | 0.25023 | 0.15399 | 2.6407 | 0.1042 | 1.284 |

**Figure 6.21.** *Results from the PHREG procedure, parameter estimation*

The program below will import the Stata format data (*uis.dta* [HOS 08])
into SAS and perform a Cox multiple regression model.

```
PROC IMPORT  OUT= WORK.UIS
DATAFILE=  "C:\folder\subfolder\uis.dta" DBMS=STATA  REPLACE;
RUN;
PROC PHREG data=uis;
MODEL time*censor(0)=age treat;
run;
```

Here the independent variables, age and treat are assumed as quantitative.

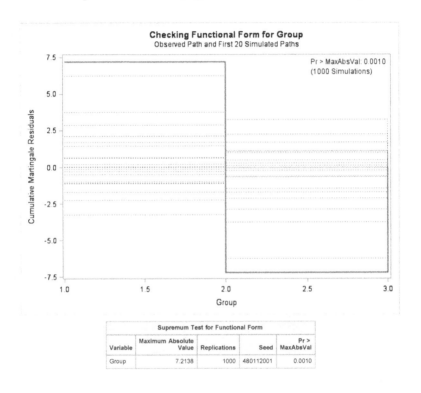

Figure 6.22. *Continuation of the results from the PHREG procedure, functional relationship test*

In the following, treat is assumed as categorical and goodness-of-fit tests are performed.

```
PROC PHREG data=uis;
CLASS treat
MODEL time*censor(0)=age treat;
ASSESS VAR=(age) PH / RESAMPLE ; run;
run;
```

First, the functional relationship for the quantitative variable age is tested (Figure 6.24).

Then, the proportional hazards assumption is tested for the two independent variables.

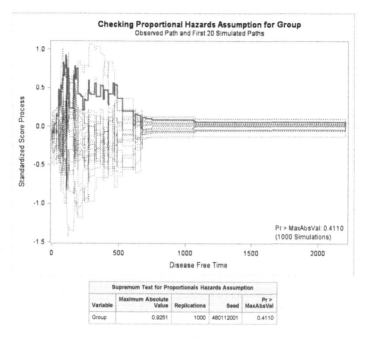

**Figure 6.23.** *Continuation of the results from the PHREG procedure, proportional hazards test*

| Analysis of Maximum Likelihood Estimates | | | | | | | |
|---|---|---|---|---|---|---|---|
| Parameter | DF | Parameter Estimate | Standard Error | Chi-Square | Pr > ChiSq | Hazard Ratio | Label |
| age | 1 | -0.01327 | 0.00721 | 3.3847 | 0.0658 | 0.987 | Age at Enrollment |
| treat | 1 | -0.22298 | 0.08933 | 6.2307 | 0.0126 | 0.800 | Treatment Randomization Assignment |

**Figure 6.24.** *Results from the PHREG procedure, two independent variables*

## 6.3. Key points to remember

In this chapter, we have essentially presented the PROC LIFETEST and PROC PHREG procedures:

– the estimate of the survival curve, known as Kaplan-Meier curve, can be obtained with PROC LIFETEST;

– it also allows the comparison between several survival curves by means of log-rank type tests;

– it produces graphics by default or by using specific options (survival curves, cumulative hazards curves);

– PROC PHREG allows for the estimation of the parameters of the Cox regression model and testing the proportional hazards assumption.

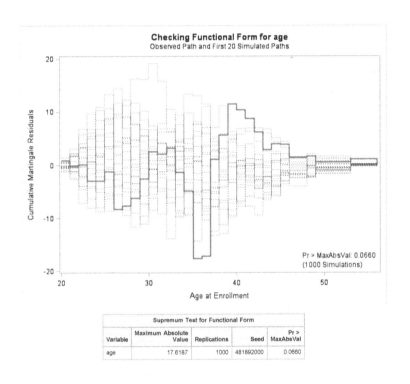

**Figure 6.25.** *Results from the PHREG procedure, functional relationship test*

## 6.4. Further Information

The PROC LIFEREG procedure for addressing parametric survival models has not been considered in this book (LIFETEST and PHREG are essentially dedicated to semi-parametric or non-semi-parametric models). For a more in-depth analysis of survival analysis, see the book by Royston and Lambert [ROY 11] or the book by Hosmer and Lemeshow [HOS 08].

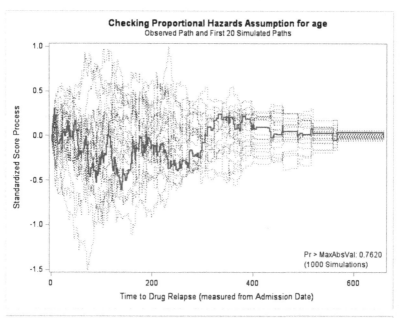

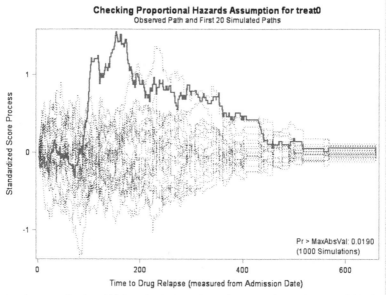

**Figure 6.26.** *Results from the PHREG procedure,
proportional hazards tests*

## 6.5. Applications

1) In a placebo-controlled trial on biliary cirrhosis, D-penicillamine (DPCA) has been introduced in the active arm in a cohort of 312 patients. In total, 154 patients have been randomized in the active arm (variable treatment, rx, 1=Placebo, 2=DPCA). A data set comprising age, biological data and varied clinical signs including the level of bilirubin serum (bilirub) are available in the pbc.txt file [VIT 05]. The patient's status is stored in the variable status (0=alive, 1=deceased) and the follow-up time (years) represents the time in years elapsed since the date of the diagnosis.

a) How many deceased individuals can be identified? What proportion of these deaths can be found in the active arm?

b) Display the distribution of follow-up time of the 312 patients, by distinctively bringing forward the deceased individuals. Calculate the follow-up median time (in years) for each of the two treatment groups. How many positive events are there beyond 10.5 years and what is the gender of these patients?

c) The 19 patients whose number (number) can be found in the following list have undergone a transplant during the follow-up period.

5 105 111 120 125 158 183 241 246 247 254 263 264 265 274 288 291 295 297 345 361 362 375 380 383

Indicate their mean age, the distribution according to sex and the median duration of the follow-up in days until transplant.

d) Display a table summarizing the distribution of number of subjects at risk according to time, with the associated survival value.

e) Print the Kaplan-Meier curve with a 95% confidence interval, regardless of the type of treatment.

f) Calculate the survival median and its 95% confidence interval for each group and display the corresponding survival curves.

g) Carry out the log-rank test considering the factor rx as the predictor. Compare with the Wilcoxon test.

h) Perform the log-rank test on the interest factor (rx) by stratifying on the age. Three age groups will be considered: 40 years old or less, between 40 and 55 inclusive and more than 55 years old.

i) Find the same results as in (g) with a Cox regression.

The data file is a text file with tabs as the field separator. It can be imported in SAS by employing an IMPORT procedure and the option DBMS, making it possible to specify the type of field delimiter used for this data file.

```
PROC IMPORT  OUT= WORK.PBC
DATAFILE=  "C:\data\pbc.txt" DBMS=TAB  REPLACE;
GETNAMES=YES;
DATAROW=2;
RUN;
PROC CONTENTS DATA=pbc;  RUN;
PROC FORMAT;
VALUE treat  1="Placebo"  2="DPCA";
VALUE Stat  0="Alive"  1="Deceased";
RUN;
```

It is possible to verify the number and the proportion of patients who died (status, 0=alive and 1=deceased) and their distribution according to the treatment group by means of simple and cross tabulations. Relative and absolute row or column frequencies are directly provided by SAS.

```
PROC FREQ  DATA=pbc;  TABLES  status*rx;
FORMAT  status  stat.  rx  treat.;  RUN;
```

In order to view the distribution of follow-up times, a simple scatter plot will be used as in the R language. To distinctly bring forward observations according to status (0 or 1), a group option will be added to the SGPLOT procedure.

```
PROC SGPLOT  DATA=pbc;
SCATTER x=years  y=number
 / group=status; FORMAT  status  stat. ;
RUN;
```

The median of the follow-up time per treatment group can be obtained by using PROC SORT and proceeding per group with the BY option.

```
PROC SORT  DATA=pbc;  BY rx; RUN;
PROC SUMMARY  DATA=pbc  PRINT  median;  VAR years;
BY rx;  FORMAT  rx  treat.;
RUN;
```

The number of deaths recorded beyond 10.5 years of follow-up is obtained with a simple one-way tabulation (PROC FREQ); however, it is necessary to create a binary variable for encoding the condition "years of follow-up > 10.5 or not".

```
DATA pbc ; SET  pbc;
EVE=0;  if years  GE  10.5  then  EVE=1; RUN;
PROC FORMAT;
VALUE pos 0="Follow-up  <  10.5" 1="Follow-up  >=  10.5";
RUN;
PROC FREQ  DATA=pbc;  TABLES  EVE*status;
FORMAT  eve pos.  status  stat.;  RUN;
```

Regarding the analysis of transplanted patients, the simplest is to create a new data table.

```
DATA transplant; INPUT number @@; tranpl=1;
cards;
5  105  111  120  125  158  183  241  246  247  254
263  264  265  274  288  291  295  297  345  361
362  375  380  383
; RUN;
```

Then, the desirable statistics are simply obtained with a combination of the MEANS, FREQ and SUMMARY procedures.

```
PROC SORT  DATA=transplant;  BY number;  RUN;
DATA all;  MERGE  pbc  transplant;  BY number;  RUN;
DATA pbc_transplant;
SET  all;  years_J=years*365; if tranpl=1;
if sample=1;
RUN;

PROC MEANS  DATA=pbc_transplant;
VAR age;  RUN; PROC FREQ  DATA=pbc_transplant;
TABLES  sex;
RUN;
PROC SUMMARY  DATA=pbc_transplant  PRINT  median;
VAR years_J;
RUN;
```

The mortality table with the estimator of the Kaplan-Meier survival function is obtained through the PROC LIFETEST procedure.

```
PROC LIFETEST  DATA=pbc;
time   years*status(0);
RUN;
```

The survival curve is obtained by adding a plot option to the previous command.

```
PROC LIFETEST  DATA=pbc plot=survival(cl);;
time   years*status(0);
RUN;
```

The analysis stratified per treatment group follows the same principle by simply adding a STRATA statement during the call to the PROC LIFETEST command, as shown below.

```
PROC LIFETEST  DATA=pbc PLOT=survival(cl);
time   years*status(0);
strata   rx;
RUN;
```

Concerning the log-rank test, it is provided by the same PROC LIFETEST command by adding the option test followed by factor.

```
PROC LIFETEST  DATA=pbc  PLOT=survival(cl);
time   years*status(0);
test   rx;
RUN;
```

By default, SAS displays the result of the log-rank test, but also that of the Gehan-Wilcoxon test.

To recode the variable age into a three-class qualitative variable, a simple DATA step can be used, as described hereafter.

```
DATA pbc;  SET   pbc;
age_class=1;
IF   age GT   40    THEN age_class=2;
IF   age GT   55    THEN age_class=3;
PROC PRINT;  VAR age age_class;
RUN;
```

The PROC PRINT command is a means to verify that the numeric values have been correctly associated with the right class for the new age_class variable. The log-rank test is obtained by using the same principle as the one exposed above, that is by adding a test statement in PROC LIFETEST followed a STRATA statement to indicate the stratification factor.

Finally, the Cox model is achieved by entering the PROC PHREG procedure. As in the case of the other regression models, the MODEL statement makes it possible to specify the relationship between the response variable (here, survival data represented by the statement years*status(0)) and the explanatory variable(s), rx in this case. In the model below, the age_class stratification variable is not included.

```
PROC PHREG data=pbc; MODEL years*status(0) = rx; RUN;
```

2) In a randomized trial, the objective was to compare two treatments for prostate cancer. The patients were orally taking every day either 1 mg of diethylstilbestrol (DES, active arms) or a placebo, and the survival time was measured in months [COL 94]. The question of interest is to know whether the survival is different between the two groups of patients, and the other variables present in the data file prostate.dat will be ignored.

a) Calculate the survival median for all patients and per treatment group.

b) What is the difference between the proportions of survival in both groups at 50 months?

c) Display the survival curves for the two groups of patients.

d) Perform the log-rank test to test the hypothesis according to which the DES treatment has a positive effect on the survival of patients.

To import the data, it is important to properly indicate the type of field separator in the import procedure. Here, we are dealing with spaces; therefore, it is necessary to indicate DELIMITER=' 'x.

```
PROC IMPORT  OUT= WORK.Prostate
DATAFILE=  "C:\data\prostate.dat" DBMS=DLM  REPLACE;
DELIMITER=' 'x; GETNAMES=YES;
DATAROW=2;
RUN;
```

The rest of the commands are roughly comparable to those used in the previous exercise.

The PROC LIFETEST command enables an estimate of the survival function to be provided by specifying via the TIME option the information necessary to identify the period (time) variables and the status (status).

```
PROC LIFETEST  DATA=prostate; TIME  time*status(0);
RUN;
```

The use of PROC PHREG below makes it possible to find in the file Diff1 the estimate of the difference between the survival proportions in both groups at 50 months. Proc print, which follows, enables its displaying.

```
proc phreg data=prostate plots(overlay)=survival;
   class treatment;
   model Time*status(0)=treatment;
   baseline outdiff=Diff1 survival=_all_/diradj group=treatment;
run;
proc print data=Diff1(where=(Time=50));run;
```

In the presence of a stratification variable, the STRATA option is added followed by the name of the stratification variable.

```
PROC LIFETEST  DATA=prostate; TIME  time*status(0);
STRATA treatment;
RUN;
```

Finally, to display the survival curve, the PLOT=survival(cl) option is added.

```
PROC LIFETEST  DATA=prostate  plot=survival(cl)
; TIME  time*status  (0);
STRATA treatment;
RUN;
```

# Appendices

# Appendix A

## Introduction to SAS Studio

SAS Studio (or SAS University Edition) is a free version of the SAS software program that can be downloaded and installed by anyone.

To whom is intended SAS Studio? SAS's answer on the dedicated website (see below) is clear. We reproduce it as it is:

– SAS University Edition *is designed for those who want easy access to statistical software to learn and perform quantitative analysis. This includes teachers, professors, students, academic researchers and independent learners.*

The Internet address of the website for information and download of this version of SAS is:

$$www.sas.com/en\_us/software/university - edition.html$$

### A.1. Dialogue with Dylan to install SAS Studio

The following reproduces a conversation with Dylan, a virtual interlocutor ("robot") from SAS Viya, to obtain information about SAS Studio in a detailed manner:

– Info: at 14:29:20 Please wait, an operator will answer you.

– Info: at 14:29:22 You are now talking to "Dylan".

– You: at 15:26:40 Hello Dylan.

– You: at 15:27:37 Would it be possible to have more information about SAS Studio?

– Dylan: at 14:32:57 Hello. What is your context?

– You: at 15:30:56 I'm a teacher and I want my students to use it.

– Dylan: at 14:35:14 Well. What University do you represent?

– You: at 15:33:46 I represent myself, teacher–researcher.

– Dylan: at 14:38:27 We have a free version which is SAS University Edition for teachers and students.

– Dylan: at 14:38:55 Link: http://www.sas.com/en_us/software/university-edition.html

– Dylan: at 14:39:40 It includes SAS Stat/SAS Base/SAS IMl and SAS Access.

– Dylan: at 14:40:05 These are the elements of SAS Studio.

– You: at 15:39:57 Can it easily be installed on Windows laptop PCs? Apple?

– Dylan: at 14:43:10 Yes of course.

– You: at 15:41:33 Do we need to be connected to the Internet by Wifi to install it? To make it work?

– Dylan: at 14:47:01 You can either install it locally or use it in the cloud.

– Dylan: at 14:47:37 For a local installation, simply download it and install it without having Internet.

– You: at 15:47:14 I'm sorry, I did not quite understand. What does use it in the cloud mean?

– Dylan: at 14:50:36 In the cloud means that you can use the software without the need to install it.

– Dylan: at 14:51:26 But an internet connection will be necessary.

– You: at 15:51:15 Thank you. If I install it locally, and so if I don't want to or cannot use it in the cloud, what are my hardware constraints?

– Dylan: at 14:55:33 For a local installation, there are no constraints, in the cloud the constraint is to have Internet access.

– You: at 15:54:54 I'm sorry, you did not understand me. I mean: is there no hardware constraint with Windows PC or Apple?

– Dylan: at 14:59:07 Here are the system requirements in Windows: Windows

– Dylan: at 14:59:07 Microsoft Windows version 7, 8, 8.1 or 10

– Dylan: at 14:59:07 64-bit hardware with a minimum of 1GB of RAM, one of the following virtualization software packages: VMware Player 7 or later Oracle Virtual Box 4.3.16 or later One of the following web browsers: Microsoft Internet Explorer 9, 10 or 11 Mozilla Firefox 21 or later Google Chrome 27 or later; with Mac: OS X ; Mac OS X 10.8 or later

– Dylan: at 14:59:22 64-bit hardware with a minimum of 1GB of RAM

– Dylan: at 14:59:22 One of the following virtualization software packages:

– Dylan: at 14:59:22 VMware Fusion for OS X 7 or later

– Dylan: at 14:59:22 Oracle Virtual Box 4.3.16 or later

– Dylan: at 14:59:22 One of the following web browsers:

– Dylan: at 14:59:22 Apple Safari 6.0 or later

– Dylan: at 14:59:22 Mozilla Firefox 21 or later

– Dylan: at 14:59:22 Google Chrome 27 or later

– Dylan: at 15:00:02 These are the only constraints for operating on PC or Mac.

– You: at 16:03:01 Thank you, I will get back to you if I have any problem to install it and to use it locally ...

– Dylan: at 15:07:29 Do not hesitate to contact us, if necessary. Good afternoon.

## A.2. Comments

– *SAS studio* can be downloaded and used by any individual (not a company).

– A valid *e-mail address* is sufficient.

– It can work in PC (Windows) or Apple Environment.

– After installation, it can work without Internet access.

# Appendix B

## Introduction to SAS Macro

SAS Macro makes it possible to achieve the automatic programming of requests (Data or Proc steps).

– SAS MACRO *is designed for SAS programmers or analysts who need to generalize their programs or improve programming efficiency.*

The Internet address of the website for documentation and download of this version of SAS is:

$$www.sas.com/en\_us/software/university - edition.html$$

### B.1. Simple examples of SAS/MACRO programs

In Chapter 5 (Logistic Regression), we executed the program hereafter in which the response variable is low and the independent variable is smoke. These two variables are both dichotomous, taking the values 0 or 1.

```
PROC LOGISTIC data=birthwt DESCENDING;
MODEL low = smoke;
run;
```

Suppose that we want to automate the execution of this program, with instead of the variable smoke, one of the variables ptl, ht, ui or ftw. We will start by renaming, in a data step, the variables smoke, ptl, ht, ui or ftw into V1, V2, V3, V4 and V5 respectively.

```
data new;set birthwt;
V1=smoke; V2=ptl;V3=ht; V4=ui; V5=ftw;
run;

OPTIONS MPRINT MLOGIC ;
%MACRO logitauto(num);
%DO i = 1 %TO &num ;
TITLE "the independent variable is V&i";
PROC LOGISTIC DATA=new DESCENDING ;
MODEL low = V&i ;
RUN ;
%END ;
%MEND ;
%logitauto(5)
```

The first line of this program, the MPRINT and MLOGIC options, produce information in the log that can help us to debug our macro. The macro itself begins from the second line ($\%MACRO$...) and is named logitauto. It contains a parameter "num" representing here the number of independent variables. It ends at the penultimate line ($\%MEND$)

The last line ($\%logitauto(5)$) orders the execution of the MACRO logitauto with the "num" parameter equal to 5.

## B.2. Comments

– SAS MACROs operate as the functions of the R language.

– The SAS MACRO Reference language provides a detailed description of MACRO facilities, and is available for download from: http://support.sas.com/documentation/cdl/en/mcrolref/69726/PDF/default/mcrolref.pdf

# Appendix C

## Introduction to SAS IML

SAS/IML (IML for Interactive Matrix Language) is a matrix programming language. It takes the form of a SAS procedure.

In the reference guide of this SAS procedure, SAS/IML is presented as it follows (we reproduce it as such):

– SAS/IML: *software gives you access to a powerful and flexible programming language in a dynamic, interactive environment. The acronym IML stands for Interactive Matrix Language.*

– *The fundamental object of the language is a data matrix. You can use SAS IML software interactively (at the statement level) to see results immediately, or you can submit blocks of statements or an entire program.*

– *You can also encapsulate a series of statements by defining a module. You can call the module later to execute all of the statements in the module.*

### C.1. Example of a SAS/IML program

– Correlation matrix computation.

– proc iml;

– /* Module for computing a correlation matrix */

– start corr;

– n = nrow(x); /* number of observations */

– sum = x[+,] ; /* computation of the columns total */

– xpx = t(x)*x-t(sum)*sum/n; /* computation of the SSCP matrix */

– s = diag(1/sqrt(vecdiag(xpx))); /* matrix normalization */

– corr = s*xpx*s; /* correlation matrix */

– print "Correlation Matrix",,corr[rowname=nm colname=nm] ;

– finish corr;

– / * Module to normalize data * /.

– start std;

– mean = x[+,] /n; /* computation of column means */

– x = x-repeat(mean,n,1); /* center x to mean zero */

– ss = x[♯♯,] ; /* sum of the column squares */

– std = sqrt(ss/(n-1)); /* estimate of the standard deviation */

– x = x*diag(1/std); /* normalization */

– print ,"Normalized Data",,X[colname=nm] ;

– finish std;

– /* Example data */

– x =  1 2 3,

– 3 2 1,

– 4 2 1,

– 0 4 1,

– 24 1 0,

– 1 3 8;

– nm=age weight height;

– run corr;

– run std;

## C.2. Comments

– SAS/IML is a SAS procedure that operates as the R language.

– The SAS/IML User's guide provides a detailed description of SAS/IML facilities, and is available for download from: https://support.sas.com/ documentation/cdl/en/imlug/65547/PDF/default/imlug.pdf

# Bibliography

[BLI 52]  BLISS C., *The Statistics of Bioassay*, Academic Press, New York, 1952.

[BRE 80]  BRESLOW N., DAY N., *Statistical Methods in Cancer Research: Vol. 1, The Analysis of Case-Control Studies*, IARC Scientific Publications, Lyon, 1980.

[COL 94]  COLLETT D., *Modelling Survival Data in Medical Research*, Chapman & Hall/CRC, Boca Raton, 1994.

[DEC 11]  DECOURT O., *SAS L'essentiel*, Dunod, Paris, 2011.

[DEL 88]  DELONG E.R., DELONG D.M., CLARKE-PEARSON D.L., "Comparing the areas under two or more correlated receiver operating characteristic curves: a nonparametric approach", *Biometrics*, vol. 44, pp. 837–845, 1988.

[DUP 09]  DUPONT W., *Statistical Modeling for Biomedical Researchers*, 2nd ed., Cambridge University Press, Cambridge, 2009.

[EVE 01]  EVERITT B., RABE-HESKETH S., *Analyzing Medical Data Using S-PLUS*, Springer, New York, 2001.

[HAN 93]  HAND D., DALY F., MCCONWAY K. *et al.* (eds.), *A Handbook of Small Data Sets*, Chapman & Hall/CRC, Boca Raton, 1993.

[HAW 99]  HAWORTH L., *PROC TABULATE by Example*, SAS Publishing, 1999.

[HOS 89]  HOSMER D., LEMESHOW S., *Applied Logistic Regression*, John Wiley & Sons, New York, 1989.

[HOS 08]  HOSMER D., LEMESHOW S., *Applied Survival Analysis*, John Wiley & Sons, New York, 2008.

[LAL 16]  LALANNE C., MESBAH M., *Biostatistics and Computer-Based Analysis of Health Data Using R*, ISTE Ltd, London and Elsevier, Oxford, 2016.

[PEA 05]  PEAT J., BARTON B., *Medical Statistics: A Guide to Data Analysis and Critical Appraisal*, 2nd ed., John Wiley & Sons, New York, 2005.

[RIN 14] RINGUEDE S., *SAS Introduction au décisionnel: du data management au reporting*, 3rd ed., Pearson Education, Paris, 2014.

[ROY 11] ROYSTON P., LAMBERT P., *Flexible Parametric Survival Analysis Using Stata: Beyond the Cox Model*, Stata Press, College Station, 2011.

[SAS 15] SAS INSTITUTE, *Step-by-Step Programming with Base SAS R 9.4*, 2nd ed., SAS Publishing, 2015.

[SEL 98] SELVIN S., *Modern Applied Biostatistical Methods Using S-PLUS*, Oxford University Press, New York, 1998.

[STU 08] STUDENT, "The probable error of a mean", *Biometrika*, vol. 6, no. 1, pp. 1–25, 1908.

[TUF 11] TUFFERY S., *Data Mining and Statistics for Decision Making*, Wiley, 2011.

[VIT 05] VITTINGHOFF E., GLIDDEN D., SHIBOSKI S., MCCULLOCH C., *Regression Methods in Biostatistics. Linear, Logistic, Survival, and Repeated Measures Models*, Springer, New York, 2005.

# Index

$\alpha = 0.05$, 105, 108, 112
$\chi^2$ test, 105
*cc_oesophage.csv*, 112
(CL) = (survival cumhaz) option, 126

## A, B, C

absolute value: ABS, 10
Access, 6
active arm, 134, 138
acute
    lymphoblastic leukemia, 116
    myelocytic leukemia, 116
ADJUST = SIDAK option, 120, 121
adjusted odds ratio, 105
age at diagnosis, 70
alcohol consumption, 112
alleles, 70
ALPHA option, 80, 81
alphanumeric type, 13
Analysis
    of covariance, 57, 80, 88, 91
    of variance table, 74, 75, 93
anthropometric characteristics, 89
ascorbic acid, 74
BACKWARD, 86
Bartlett test, 74
baseline, 139
basic operator, 3
benign breast disease, 68
bilateral test, 67

biliary cirrhosis, 134
biological
    data, 134
    measurements, 15
birth, 65, 66, 94, 115
birthwt.txt, 7
Bivariate descriptive statistics, 29–32
blood pressure, 91–93, 108, 109, 111, 112
body mass index, 91, 92
BON option, 61, 65
Bone marrow transplant, 116
Bonferroni, 59, 71
    method, 71
box plots, 40, 64, 72
Bravais-Pearson's coefficient, 90
breast cancer, 70, 127
BY
    instruction, 13
    statement, 13, 21–23, 71
C vitamin, 74
CARDS, 52
censored time, 115
censoring, 115, 116
    indicator, 116
CHISQ option, 29, 64, 106
Chi-Square independence test, 64
CHI-SQUARE test, 29, 30, 48–62, 69
Chi-Square test with continuity correction, 48
cholesterol rate, 108
CLASS statement, 57, 67, 71, 96, 126

Printed in the United States
By Bookmasters